A revolution in
the Earth sciences

A. HALLAM
FELLOW OF NEW COLLEGE, OXFORD

A revolution in the Earth sciences

FROM CONTINENTAL DRIFT TO PLATE TECTONICS

CLARENDON PRESS · OXFORD

Oxford University Press, Ely House, London, W.1

GLASGOW NEW YORK TORONTO MELBOURNE WELLINGTON
CAPE TOWN IBADAN NAIROBI DAR ES SALAAM LUSAKA ADDIS ABABA
DELHI BOMBAY CALCUTTA MADRAS KARACHI LAHORE DACCA
KUALA LUMPUR SINGAPORE HONG KONG TOKYO

Printed in the United States of America

Preface

THE LAST few years have been an extremely exciting time to be in the Earth sciences, just as they were a decade earlier in molecular biology. Bold new ideas bearing on fundamental aspects of the structure and evolution of our planet have been put to test in various ways and confirmed to general satisfaction. Much of our thinking has had to undergo severe reappraisal, agonizing or otherwise, and application of the new concepts to a more accurate and complete interpretation of geological history is still in a youthful phase.

Now that these spectacular modern developments are just beginning to be noted in the textbooks, and courses in the more progressive universities are undergoing some necessary reorientation, the time seems ripe to examine them in their historical context, which is Wegener's hypothesis of continental drift. The immediate stimulus to writing this book comes from John Ziman. I do not know of a better account of what scientists do, as opposed to what they are commonly supposed to do, than his book *Public knowledge: the social dimensions of science*.

It is reassuring to find a distinguished physicist confirming one's suspicion that physics is a rather atypical science in many respects, contrary to the views of those by no means rare philosophers who tend to treat physics and science as virtually synonymous. I believe, with Ziman, that those interested in the history of science may learn valuable lessons from the story of the continental drift controversy and its sequel. Few subjects can better illustrate the interplay of fact and hypothesis, of competing schools of thought and personality conflicts, of the pervasive influence of intellectual fashion, and of what it takes to achieve a consensus of working scientists.

The subject lends itself well to the narrative form and hence this approach has been adopted here. At the same time I hope that at the end of the story the reader may have a reasonably comprehensive idea of the modern geology in so far as it bears on some of the most important questions that have been posed since the early days of the science.

In aiming the book at essentially two groups of people, those either undergoing or having completed professional training in the Earth sciences, and laymen interested in the history of science, I am well aware of the dangers. The challenge is to steer a middle course as best one may between the Scylla of oversimplification that alienates the *cognoscenti* and the Charybdis of technical exposition that confuses the rest. Fortunately the basic concepts are quite simple and can be presented with a minimum of jargon. Little prior geological knowledge is assumed, but rather than include a glossary I have attempted brief definitions within the text of some of the less familiar terms. The geological time-scale for the last demi-eon is given in an appendix because a knowledge of the succession of geological periods and their relative length is vital to understanding.

Although the layman can appreciate the main points of the story without special training there is no question that consultation of one of the many elementary textbooks of geology would considerably aid his understanding and give him a better perspective. We need only remember that there are *degrees* of understanding, that one may still get a better view of the landscape only halfway up the hill than at the foot.

As for those who already know some geology, and who may even be well acquainted with plate tectonics and its ramifications, it is hoped that the historical treatment may prove at least to some degree illuminating. Sometimes when I hear certain professional colleagues talk of Wegener's 'classic' book they help to confirm my private definition of a classic as a work everyone quotes but no one actually bothers to read. How many currently active research workers or students have even read Hess on sea-floor spreading, let alone Holmes's anticipatory 1929 paper?

It is often maintained that, such is the difficulty of coping with the flood of current literature, the researcher who wishes to stay abreast of new developments simply lacks the time for the more scholarly pursuit of probing into the history of his subject. I believe on the contrary that such an attempt is well worth the effort, because the loss in time reading the current journals may be more than compensated by the gain in perspective and insight into the nature of his research.

The sheer volume of scientific literature, in this field as in any other, is of course a bugbear. It would have been easy to write a book twice the length by going into fuller detail or quoting more published work, but I desisted because of a reluctance to indulge in what I would have considered needless inflation. Instead I have tried to tell the story as I see it quite sparingly, but with enough references at the end of the book to allow the interested reader to follow up matters in greater depth over the whole range of the subject matter. Those cited are restricted to key works, reviews, or papers that enlarge on specific points made in the text.

I am indebted to Jim Briden, Steve Gould, Alan Smith, Fred Vine, and Rom Harré for critically reading various chapters, and to many more friends and colleagues who have helped to mould my opinions and improve my knowledge over the years, but I naturally take full responsibility for the views and presentation of what follows.

Oxford   A. H.
June 1972

Acknowledgements

THANKS are due to the following publishers and journals and to the authors concerned (who are acknowledged in the underlines) for permission to reproduce the illustrations listed below.

The Geological Society of America, Fig. 1; Messrs Methuen and Co., Figs. 2, 3, 4, 5; *Scottish Journal of Geology*, Fig. 10; Messrs Oliver and Boyd, Figs. 11, 12; the Open University, Fig. 13; the American Association for the Advancement of Science, Figs. 20, 26, 39; *Nature*, Figs. 23 and 25 and Appendix 2; Messrs W. H. Freeman, Figs. 27, 28; the American Geophysical Union, Figs. 30, 33, 36, 37; Messrs Blackwells Scientific Publications Ltd., Fig. 40; *American Journal of Science*, Figs. 42, 43.

Contents

A revolution in
the Earth sciences

1. Wegener's precursors

One curious result of this inertia, which deserves to rank among the fundamental 'laws' of nature, is that when a discovery has finally won tardy recognition it is usually found to have been anticipated, often with cogent reasons and in great detail.

F. C. S. SCHILLER

ALTHOUGH the hypothesis of continental drift is quite generally and rightly associated with the name of Alfred Wegener, he was not the first person to suggest that the continents had moved apart. There has been controversy, however, about who this first person actually was.

Francis Bacon appears to be the earliest writer to whom the germ of the idea has been attributed. Clearly, an essential prerequisite for any ideas concerning the relationships of continents and oceans is a reasonably comprehensive world map. Such existed of sufficient accuracy to make meaningful speculations possible by 1620, when Bacon wrote his *Novum Organum*. An examination of his writings shows that, far from hinting at a splitting apart of Africa and South America, he merely pointed out a general conformity of outline of the two continents, both of which tapered southwards, and a certain similarity between the Pacific coast of South America and the Atlantic coast of Africa.

Equally misplaced is the attribution of the idea to the French moralist François Placet, who in 1666 published a booklet entitled *La Corruption du grand et petit Monde*. Amidst various fantastic ideas, he proposed that before the Noachian flood the Earth must have been undivided. The separation of America was not, however, accomplished by drifting but either by depression of 'Atlantis' and concomitant uplift of a continent to the west, or by an agglomeration of islands.

Subsidence of a former land, 'Atlantis', on the site of the present Atlantic Ocean, also played a role in the conjectures of the Comte de Buffon. At the beginning of the nineteenth century the German explorer, Alexander von Humboldt, was struck by the congruence of the eastern South American and west African coasts and speculated that the Atlantic was nothing more than a huge valley scooped out by the sea. Once more there was no suggestion of an actual moving apart of the bordering continents.

It is not until the publication in 1858 of *La Création et ses mystères dévoilés* by Antonio Snider-Pellegrini that we obtain the first clear indication of a break-up and drifting apart of the Atlantic continents. Snider was an old-fashioned catastrophist who speculated that during the cooling and crystallization of the Earth from a molten mass the continents had formed only on one side, thereby creating a condition of instability which was relieved during the Noachian flood only by extensive catastrophic fracturing and pulling apart of the Americas from the Old World. The coastal fit of Africa and South America was explicitly cited as supporting evidence (Fig. 1).

FIG. 1. Snider's interpretation of the position of the continents before the opening of the Atlantic.

With the progressive spread of influence of Lyell's uniformitarian doctrines during the latter part of the nineteenth century there was little chance of Snider's evidently fantastic notions being considered seriously by the scientific community. Probably to its detriment, the idea of continental drift continued to be associated with a catastrophic event. Following G. H. (later Sir George) Darwin's proposal in 1879 that the Moon had been born out of the Earth at an early stage in the latter's history, leaving behind the gigantic scar of the Pacific, Osmond Fisher indicated that one likely consequence would have been the lateral movement and fragmentation of the cooled granitic crust. Fisher's comprehensive geophysical treatise entitled *Physics of the Earth's crust*, apparently the first of its kind, makes stimulating reading, not least for his postulation that the Earth's relatively fluid interior was subjected to convection currents rising beneath the oceans and falling beneath the continents—a remarkable anticipation of later ideas.

The idea of some association between the departure of the Moon from the Pacific and lateral movement of continental masses persisted into this century, for instance in the writing of W. H. Pickering and H. B. Baker. The latter even suggested an age as young as the Tertiary for the event. Fisher's mobilist views were not typical of what might be called the Anglo-American geophysical tradition in the late nineteenth and early twentieth centuries, which laid emphasis on the properties of the solid Earth and appeared to deny the possibility of, for instance, migration of the north and south poles. They accorded better with the contemporary Germanic tradition, which because of the language barrier was not well understood in the English-speaking world. Scientists such as Wettstein, Loeffelholz von Colberg, and Kreichgauer integrated meteorological and climatological data fully into geophysics, unlike their British and American colleagues, and adopted mobilist views, with segments of the Earth's crust floating on a liquid interior. No dramatic leap of the imagination was required therefore to envisage the possibility of polar wandering.

The importance of this Germanic tradition has, of course, its main significance in having provided a matrix of ideas which must strongly have influenced Wegener's thinking, as we shall see in the next chapter. It was, however, an American scientist, F. B. Taylor, who in 1910 published a lengthy paper giving the first logically worked out and coherent hypothesis involving what we would now recognize as continental drift.

The starting-point of Taylor's hypothesis is not, as one might expect, the supposed fit of the continents bordering the Atlantic but the pattern of Tertiary mountain belts in Eurasia. Taylor had evidently been highly stimulated by the comprehensive description given of these belts by Eduard Suess in *The face of the Earth*. On the eastern and southern borders of Asia and continuing into the Mediterranean region are a series of arcuate zones of mountains, usually convex towards the ocean, which exhibit signs of lateral compression in the form of folded and overthrust strata. From his analysis of the trend lines of these ranges Suess had concluded that they could be interpreted as the result of oceanic depression and tangential thrusts directed towards the ocean from certain northern vertices or horsts, consequential upon cooling contraction of the Earth.

Taylor found the conventional contraction hypothesis inadequate to explain satisfactorily the distribution and youth of the Tertiary mountain ranges. He envisaged instead 'a mighty creeping movement' of the Earth's crust from the north towards the periphery of Asia. The Indian peninsula, an ancient shield area, acted as an obstructive block, causing the huge pile-up of the Himalayas and Pamir plateau directly to the north, while farther east the fold ranges were able to swing round more freely, into Malaysia and Indonesia (Fig. 2). The trend lines in southern Europe are much more complicated than in Asia. Taylor ascribed this to several factors: the relative smallness of the European crustal area and hence relative feebleness of the thrust forces, the obstructing and complicating action of older ranges, tangential thrusting from the east, and resistance by the African block.

The notion of crustal creep from high to low latitudes in the northern hemisphere was supported in Taylor's paper by reference to Greenland, which was envisaged as the remnant of an old massif from which Canada and northern Europe had broken away along rifts. Suess and others had been struck by the close resemblance of Palaeozoic rocks and structures on the two sides of the North Atlantic but had attributed this to the collapse of Atlantis rather than the drifting apart of continental blocks.

Less was said by Taylor about the southern hemisphere but Australia was considered to have shifted north-eastwards, from the evidence of Tertiary fold belts in New Guinea and neighbouring areas. The Mid-Atlantic Ridge, known even at that time as a major submarine mountain range parallel to the coastal margins, was seen as the line of rifting apart of Africa and South America. Whereas, however, the latter continent had evidently moved westwards in the Tertiary, as indicated by the Cordilleran ranges, the lack of such young mountain belts on the African side was an indication of movement no later than the Carboniferous.

Certain points of detail in Taylor's interpretation are of considerable interest in the light of our modern knowledge. Thus the eastward curvature of the Scotia Arc, between Patagonia and Grahamland, was seen as an indication that westward drift of the continental masses had lagged behind in this region, signifying a minimum movement of 400 or 500 miles. Furthermore, Greenland was considered to have been displaced eastward relative to Grant, Grinnell, and Ellesmere lands of Canada along the line of what we recognize today as a transcurrent or strike-slip fault, as the Davis Strait was opened up.

Taylor paid little attention to the mechanism of continental movement in his 1910 paper but in subsequent papers suggested the operation of tidal forces when the Moon was captured, rather than lost, by the Earth during the Cretaceous. Interesting as his ideas are, they were

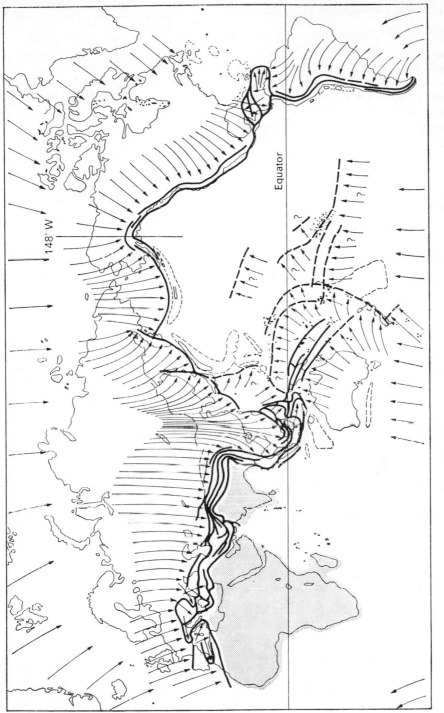

Fig. 2. Taylor's diagram to shew the direction and amount of Tertiary crustal movement.
After Taylor 1910, Fig. 7.

little supported by independent evidence. Thus he is content to record without further elaboration that 'there are many bonds of union which show that Africa and South America were formerly united'. This must be one of the chief reasons why his 1910 paper had relatively little impact on the geological community. It seems, furthermore, that Wegener developed his ideas independently only a short while afterwards. Unlike Taylor, he undertook a thorough geophysical analysis and produced over the years a considerable amount of supporting evidence from a variety of sources. In this respect the relationship of Taylor to Wegener is not unlike that of Wallace to Darwin.

2. Wegener's hypothesis

We are like a judge confronted by a defendant who declines to answer, and we must determine the truth from the circumstantial evidence. All the proofs we can muster have the deceptive character of this type of evidence. How would we assess a judge who based his decision on part of the available data only?

ALFRED WEGENER

ALFRED WEGENER was born in Berlin in 1880, the son of an evangelical preacher. He was educated at the Köllnisches Gymnasium in Berlin and subsequently at the Universities of Heidelberg, Innsbruck, and Berlin. In 1906 he went with a Danish expedition to north-east Greenland for a couple of years to undertake meteorological research. Upon his return to Germany he was appointed Privatdozent in astronomy and meteorology at the University of Marburg and wrote a well-known textbook on meteorology. In 1912 he undertook a second, rather unsuccessful Greenland expedition with J. P. Koch. After the First World War, in which he served as a junior officer and was wounded in the arm and neck, Wegener returned to academic life in Hamburg and in 1924 accepted the offer of a specially created chair in meteorology and geophysics at the University of Graz in Austria. He went on a third expedition to Greenland in 1930, where unfortunately he lost his life on the inland ice-cap.

To appreciate the revolutionary nature of Wegener's hypothesis of continental drift, it is necessary to sketch in the conventional world view of the Earth's structure and evolution that was widely held at the beginning of this century. This was eloquently described by one of geology's great synthesists, the Austrian Eduard Suess, in his multi-volume treatise *Das Antlitz der Erde*, published late in the nineteenth century and subsequently translated into English as *The face of the Earth*.

The Earth was supposed to be still in the process of progressive solidification and contraction from a molten mass. Lighter rock materials had moved towards the surface to give rise to granitic-type igneous and metamorphic rocks, with associated sediments, collectively termed *sal* (late changed to *sial*) because they were relatively rich in silicates of alumina, together with soda and potash. They were underlain by denser rocks termed *sima*, resembling, if not exactly matching, basalt, gabbro, or peridotite, which were richer in silicates of iron, calcium, and magnesium.

Mountain ranges were produced by contraction in a manner somewhat analogous to the crinkles developed on a shrinking, drying apple. On a larger scale, an overall arching pressure caused certain sectors of the Earth surface to collapse and subside, giving rise to the oceans, while the continents remained emergent as unfaulted blocks or 'horsts'. In the course of time, certain continental areas in turn sank faster than adjacent areas and hence were inundated by the sea, while temporarily stabilized parts of the ocean floor at other times emerged once again as dry land.

Evidence of former land connections across what was now deep ocean was provided abundantly by the total or near-identity of many fossil animals and plants found on different continents. Unless such trans-oceanic land bridges had existed in the past these widely acknowledged similarities in former organic life were inexplicable in terms of Darwinian evolution. Genetic isolation should have given rise to morphological divergences in the faunas of the different continents. Suess gave the name *Gondwanaland* to a former continent embracing central and southern Africa, Madagascar, and peninsular India, after the late Palaeozoic *Gondwana* fauna common to its components. The term *Gondwanaland* has in its general usage been subsequently extended to embrace Australia, South America, and Antarctica as well. *Gondwana* (originally the name of a region in eastern India) is actually the more correct term, since it means land, but Gondwanaland has probably become too familiar to be abandoned.

Suess also coined the term *eustatic* for the world-wide rises and falls of sea-level that could be inferred from the stratigraphical record of successive marine transgressions and regressions over the continents. He attributed the regressions to subsidence of the ocean basins and the transgressions to the partial filling of these basins by sediment from the continents. Water would therefore either be drained off the continents as the oceans deepened, or displaced on to them as a consequence of ocean-floor sedimentation.

Quite evidently, the interpretative model of a contracting Earth appeared to be successful in accounting for a wide range of geological phenomena. It must, indeed, have seemed more or less unassailable to many geologists early this century, for crucial tests whose interpretation was unambiguous were not conspicuously evident. Wegener thought otherwise.

According to one of his fellow-expeditioners in Greenland, Lauge Koch, the germinal idea of continental drift came to Wegener from his observing the splitting apart of slabs of ice in the sea. This rather romantic but plausible view receives no written confirmation, however, from Wegener himself.

By his own account, the first suspicion that the continents might have moved laterally came to Wegener in 1910, when he was struck by the remarkable congruence of the coastlines on either side of the Atlantic. He nevertheless initially treated the idea as improbable but the next year he came across a report of palaeontological evidence for a former land bridge between Brazil and Africa. A search for more such evidence produced in his mind such weighty corroboration that he was led on to develop the hypothesis which received its first public airing at a lecture in Frankfurt in January 1912. There followed later that year two short publications in *Petermanns Mitteilungen* and *Geologische Rundschau*, entitled 'Die Entstehung der Kontinente'. An enlarged version first appeared in book form in 1915, as *Die Entstehung der Kontinente und Ozeane*. Revised editions appeared successively in 1920, 1922, and 1929. The third edition attracted the most attention and was translated in 1924 into English, French, Spanish, and Russian, the English version appearing under the title *The origin of continents and oceans*. In this edition Wegener's expression 'Die Verschiebung der Kontinente' was quite accurately translated as 'continental displacement'. Nevertheless the subsequently coined term *continental drift* caught on universally in the English-speaking world, presumably because people would rather not utter seven syllables when five will suffice.

It is illuminating to compare the earliest published paper with the later editions of the book. In the article in *Petermanns Mitteilungen* Wegener indicates at the beginning that he is putting forward only a working hypothesis which he expects will require modifications in the course of time. Rather than outlining the 'jigsaw fit' and palaeontological clues he plunges directly into geophysical arguments, indicating what in his view are inadequacies or contradictions in orthodox theory. Only then does he cite geological evidence supporting the idea that formerly united continents have rifted and moved apart. He pinned great hope on future geodetic observations to prove that movement is still continuing, and he inferred polar wandering from the shifting of ancient climatic belts. Wegener said very little about the controlling mechanism of movement but he tentatively put forward both tidal forces and a 'flight from the poles' (*Pohlflucht*) as possibilities.

In the last edition of the book both the character of the principal arguments and the way they are marshalled remain remarkably similar, considering the passage of seventeen years. The main difference is in their greater elaboration and the addition of much new evidence, most interestingly perhaps in the field of palaeoclimatology. As befitted a meteorologist, Wegener had become increasingly interested in ancient climates, and in 1924 wrote a book on the subject in conjunction with W. Köppen. In the summarized account that follows, I shall naturally concentrate on the fourth edition of *The origin of continents and oceans* as representing the latest and most developed views that Wegener held.

The contracting Earth model as expounded by Suess was vulnerable to attack on several grounds. The then fairly recent discovery of enormous overthrusted rock slices or 'nappes' in the Alps led to estimates of Tertiary contraction which seemed excessive. Why, moreover, were not the fold 'wrinkles' more uniformly distributed over the Earth's surface, instead of being localized in narrow belts? Basic assumptions about the Earth's cooling had been undermined by the discovery of widespread radioactivity in rocks, leading to the production of considerable amounts of heat acting in opposition to the thermal loss into space by radiation.

The vast majority of the marine sedimentary rocks found on the continents are of shallow-water type and could not have been laid down in oceanic deeps. Hence the continental masses must be 'permanent' and the old notion, dating back to Lyell, that continents and oceans are in effect interchangeable could not be upheld. However, if we accept the former existence of huge land areas in sites now occupied by ocean, without the possibility of compensating for this by the submergence of present-day continental regions to the level of the ocean floor, there would be no room for the volume of the world's oceans in the now much reduced deep-sea basins. The displacement of water would in fact be so enormous that the level of the world's oceans would rise above *all* the continents. One is therefore forced into postulating subordinate *ad hoc* hypotheses, such as ocean basins at a former time deeper than today by precisely the required amount, or a substantial change in the volume of ocean water.

Furthermore, gravity data indicated that the ocean floor was underlain by denser rocks than on the continents (*sima*, in fact, as opposed to *sial*). The concept of isostasy (see below), developed on the basis of gravity observations, rendered impossible the subsidence of vast continental areas into oceanic deeps. This point had been better appreciated in America, where leading geologists such as Dana and Willis had recently argued in favour of the permanence of the ocean basins. Such an hypothesis is clearly incompatible with sunken continents and the contradiction between the rival views provides one of the starting-points for the hypothesis of horizontal displacement of the continents.

Wegener postulated that, commencing in the Mesozoic and continuing up to the present, a huge supercontinent, 'Pangaea', had rifted and the fragmented components had moved apart (Fig. 3). South America and Africa began to separate in the Cretaceous, as did North

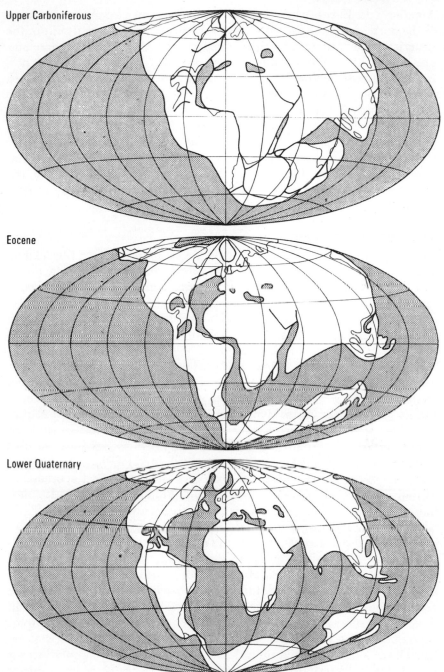

FIG. 3. Wegener's reconstruction of the various positions of the continents from the Carboniferous to the Quaternary. After Wegener 1929, Fig. 4.

America and Europe, but the last two continents had retained contact in the north as late as the Quaternary. During the westward drift of the Americas, the western Cordilleran ranges had been produced by compression at or near the leading edges, but the Antilles and the Scotia Arc had lagged behind in the Atlantic. The Indian Ocean had begun to open up in the Jurassic but the principal movements took place in the Cretaceous and Tertiary. A large area of land to the north of India had crumpled up in the path of India during its northward movement, to form the Himalayan and associated mountain ranges. Australia–New Guinea had severed its connection with Antarctica in the Eocene and moved northwards, driving into the Indonesian archipelago in the late Tertiary. The hypothesis was supported by a wide array of geophysical, geological, and biological data.

Geophysical arguments

Statistical topographic analysis of the Earth's surface reveals two predominant levels, corresponding to the continent and ocean floors. This is consistent with the notion of two separate crustal layers, a lighter sialic layer being underlain by a denser simatic layer which also forms the ocean floor. It is inconsistent with the concept of random subsidence and uplift from an original uniform level, which would lead to a normal or Gaussian distribution of levels (Fig. 4).

Before the arguments based on isostasy are outlined we must digress a little to see how the concept arose. Back in the middle of the nineteenth century geodesists were surprised to discover that mountain ranges such as the Himalayas failed to exert the gravitational pull expected from their greater mass by comparison with the adjacent plains. Of the alternative hypotheses proposed to account for this phenomenon, that of Airy became widely adopted late in the century as being the more compatible with geological data. Airy's hypothesis of isostatic compensation postulated that high mountain ranges are underlain by correspondingly deep roots of relatively low-density, sialic rock, whereas low plains were underlain by shallow roots; hence the pull of gravity over the different topographic regimes tended towards equalization. As the mountains were worn down by erosion, so the base of the sialic layer (i.e. the continental crust) would rise. Sedimentation in an adjacent structural basin would, on the other hand, tend to depress the crust because of the added load. Post-glacial uplift of Scandinavia in the past few thousand years was clear evidence of delayed isostatic uplift consequent upon the removal of the superincumbent load of Pleistocene ice. The analogy has often been drawn with blocks of wood of different height floating in a tank of water; the taller blocks rise higher out of the water. If one depresses such a block with one's finger it will bounce

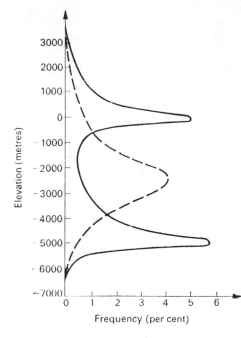

FIG. 4. The two maxima in the frequency distribution of elevations of the Earth's surface. After Wegener 1929, Fig. 8.

back to its normal 'level of compensation' immediately the applied stress is removed.

Quite clearly isostasy theory assumes that the substratum underlying the Earth's crust acts as a fluid, albeit a highly viscous one. Wegener argued that if the continental blocks can move vertically through this substratum there is no reason why they should not also be able to move horizontally, provided only that there are forces of sufficient power to do this. That such forces do in fact exist is proved by horizontal compression of strata in mountain ranges such as the Alps, Himalayas, and Andes.

Some confusion existed, Wegener pointed out, about the physical properties of the Earth. Studies of earthquakes, polar fluctuations, and tidal deformations have indicated an Earth as stiff as steel which behaves as an elastic body when acted on by short-period forces such as seismic waves. Over the much longer periods signified by geological time, however, the Earth must behave as a fluid; for instance, the oblateness of the spheroid corresponds exactly to the period of rotation. We should not be too dogmatic about the viscosity coefficient of the Earth's interior because we lack the necessary data. As early as 1912

Wegener drew the analogy with pitch, which shatters under a hammer-blow like a brittle solid, but will in the course of time flow plastically under its own weight, i.e. subject to the much milder but persistently imposed force of gravity.

By the 1920s combined gravity and petrological data, supplemented by a limited amount of seismic data, suggested that the oceans were underlain by a material denser than and fundamentally different from that of the continents. When all the available information was combined this oceanic material seemed to consist most probably of peridotite (composed of olivine and pyroxene) or dunite (olivine), or perhaps eclogite (garnet and pyroxene), with a layer of basalt on top.

Wegener placed great faith in geodetic observations by repeated astronomical position-finding or, better, by the method of radio time transmissions across lines of longitude over a number of years, to prove horizontal movement of the continents at the present day. He thought that the precision and reliability of such measurements would be such as to persuade the most sceptical scientists. By 1929 the available evidence seemed at least to indicate westward drift of Greenland with respect to Europe.

Geological arguments

Wegener gave most attention to similarities on the two sides of the Atlantic. The Cape fold belt of South Africa, for instance, appears to find its continuation in the range of Buenos Aires Province in Argentina. The ancient gneiss plateau of Africa bears many close resemblances to that of Brazil, for instance in various types of igneous rocks and kimberlite. The late Palaeozoic–early Mesozoic series of largely non-marine strata known in South Africa as the Karroo System and in Brazil as the Santa Catharina System are remarkably similar in many respects, for instance in the possession at the same stratigraphical horizon of the glacial conglomerates known as tillites. As for the sides of the North Atlantic, the early Palaeozoic ('Caledonian') and late Palaeozoic ('Armorican') mountain belts of western Europe can be traced into Newfoundland and Nova Scotia. Particular significance was attached by Wegener to the matching of the terminal moraines of the North American and European ice-sheets as indicating postponement of final rupture of the continents until the Pleistocene (Fig. 5).

In Wegener's own (translated) words, 'It is just as if we were to refit the torn pieces of a newspaper by matching their edges and then check whether the lines of print run smoothly across. If they do, there is nothing left but to conclude that the pieces were in fact joined this way. If only one line was available for the test, we would still have found a high probability for the accuracy of fit, but if we have n lines,

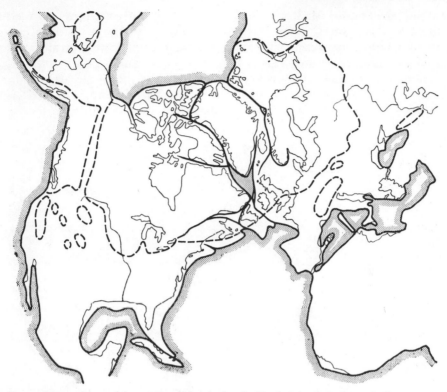

FIG. 5. The positions of the continents bordering the North Atlantic in the early Quaternary, before North America was detached. The broken lines indicate the boundaries for inland ice. After Wegener 1929, Fig. 19.

this probability is raised to the nth power.' Indeed, there is more to the celebrated jigsaw analogy than the outlines of the pieces.

Wegener said less about the opening of the Indian Ocean but he briefly referred to similarities between the different components of 'Gondwanaland' such as Madagascar, India, and Australia. He emphasized how much the drift hypothesis helps in accounting for the complicated geology of the Moluccan islands in Indonesia, whose trend lines have been bent round by the impingement of Australia–New Guinea.

Palaeontological and biological arguments

Wegener took some pains to establish that the widespread consensus among leading palaeontologists early this century was in favour of transoceanic land bridges which must have subsided into the ocean after the Cretaceous. On the basis of faunal and floral identities or similarities between different continents, land connections or (for neritic organisms) shelf-sea connections of some sort were deemed necessary by most workers. Agreement was particularly strong about the Mesozoic connections between Brazil and Africa, Australia and Africa–India, and South Africa–Madagascar and India (Fig. 6). Among the examples cited by Wegener were the small reptile *Mesosaurus*, known only from the Permian of South Africa and Brazil, and *Glossopteris*, a late Palaeozoic plant widespread in but confined to the

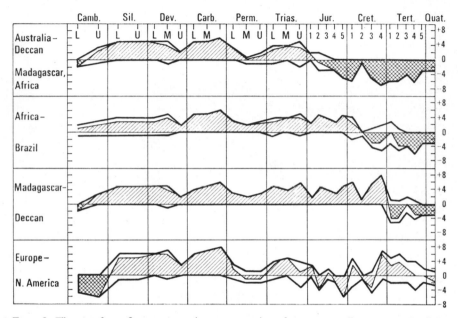

FIG. 6. The number of proponents (upper curves) and opponents (lower curves) of the existence of four land bridges since the Cambrian. The difference (majority) is hatched, and cross-hatched when the majority opposes. After Wegener 1929, Fig. 1.

southern continents, which figured importantly in the concept of Gondwanaland.

The distribution of a number of living organisms was also cited in support of former land connections, for instance earthworms and marsupials. The Australian marsupials had clearly evolved in isolation since at least the early Tertiary, but a link with South America shortly before this time was suggested by the occurrence in that continent (and absence from the Old World) of marsupial opossums. In support of this, the Australian and South American marsupials evidently share identical or at least closely similar parasites! With regard to earthworms, which of course would have the greatest possible difficulty negotiating an ocean crossing, Wegener was impressed by the data of the German zoologist Michaelson, who produced distribution maps of the lumbricid and megascolecid families to show how difficult their distribution was to explain without invoking continental drift. The maps showed strong affinities between the faunas of North America and Europe, those of South America and Africa, and those of Australia, India, South Africa, and Patagonia.

If former land connections between the continents where there is now deep ocean are an unavoidable inference, it is impossible to sink them out of sight because of the geophysical evidence and isostasy. The only reasonable conclusion to be drawn is that the now-isolated continents have shifted laterally from a formerly united supercontinent.

Palaeoclimatic arguments

At the present day we can distinguish a number of climatic zones broadly paralleling latitude. Flanking the equatorial rain belt are the high pressure arid and monsoonal belts of the Horse Latitudes, followed in turn by belts in the temperate zone with cyclonic rainfall, and finally the polar zones with ice-caps.

The most important geological evidence bearing on ancient climates is that of tillites, or glacial boulder-beds, resting on striated pavements of resistant rocks. These can be unequivocally interpreted as the trace of former ice-sheets. Coals signify humid conditions since they can of necessity form only in swampy conditions but are not good temperature-indicators. Nevertheless exceptionally thick seams of coal probably signify tropical conditions of luxuriant plant growth. Arid conditions are marked by deposits of rock-salt and gypsum, proving an excess of evaporation over precipitation, as well as by desert sandstones, which have characteristic petrological features. Judging by the present-day distribution of calcium carbonate muds and sands, and by knowledge of the decrease in solubility of calcite with rising temperature, thick limestones are likely indicators of tropical or subtropical conditions.

Fossil organisms can also prove useful as palaeoclimatic indicators. Thus the lack of annual rings in logs of wood usually signifies tropical conditions (because of the lack of seasonal contrast), and large reptiles invariably signify a warm climate. Surprisingly, reef corals are not mentioned, although they are among the best indicators of a warm climate.

Using criteria of the type outlined, it has become quite evident that, between the Carboniferous and the present, Europe's climate had changed from tropical to temperate and Spitsbergen's from sub-tropical to polar. Indications of a reverse change in South Africa, from polar to subtropical, supported the suggestion that the poles had in fact migrated gradually over a long period of time.

Equally fascinating was the distribution of Carboniferous and Permian glacial deposits in South America, South Africa, India, and Australia, the components of Gondwanaland. The best known of these, the Dwyka Tillite of South Africa, was known to be both thick and extensive. It contained far-travelled boulders and was seen in places to rest on superb striated pavements. Locally it is interbedded with marine deposits, signifying that the ice must have reached sea-level. Hence we are dealing with traces of an ice-sheet and not just a mountain glacier. Regional stratigraphy in eastern Australia had demonstrated that whereas in Tasmania and Victoria there was only one thick glacial layer, and in Queensland none, the intermediate area of New

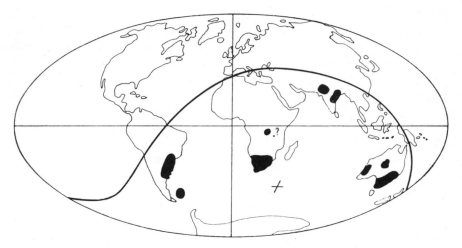

FIG. 7. The Permo-Carboniferous glaciation. The areas in black represent postulated ice-caps and the bold line the contemporary equator. After Wegener 1929, Fig. 34.

South Wales had two glacial horizons separated by coal-bearing, presumably interglacial, deposits. These observations suggested an ice centre to the south.

With the south pole at the most favourable point conceivable, about 50°S, 45°E, inland ice traces farthest from the poles (India, Brazil, eastern Australia) lie at 10°S (Fig. 7). In other words, with the continents in their present position, a polar climate would have extended almost to the equator while the other hemisphere enjoyed a tropical or subtropical climate. This is clearly absurd. How much more satisfactory if we could reassemble the continents, bringing, for instance, India a long way south from its present position north of the equator.

Confirmation that this is correct procedure comes from an examination of the Carboniferous coal measures of the eastern United States,

Europe, and China. On a reconstructed Pangaea these fall into an evidently equatorial belt, 90° from the centre of a large region of inland ice! The tropical character of the Coal Measures vegetation appears to be supported not only by the absence of tree rings but by the presence of tree-ferns with large fronds and aphlebias and by the thickness of many of the seams. The Permian coals of southern Africa contain, on the other hand, a quite different flora and seasonal growth rings are present in fossil logs. In Permian times polar wandering caused arid conditions to spread over the former equatorial zone as the climatic belts were displaced southwards. This is signified by the presence of the

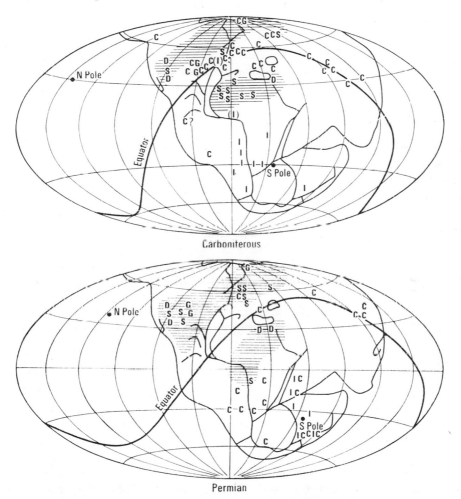

FIG. 8. Climatic belts of the Carboniferous and Permian. Hatched area is arid zone. *C*, coal; *I*, ice; *D*, desert sandstone; *S*, salt; *G*, gypsum. Modified from Wegener 1929, Figs. 35, 36.

Zechstein salt deposits in northern Europe and similar deposits in both the Urals and southern United States. Previously, in the Carboniferous, arid-zone deposits in the form of gypsum had been confined to Spitsbergen (Fig. 8).

Wegener devoted a further chapter entirely to polar wandering. 'Superficial' polar wandering can be detected only by geological evidence for climate. Complications arise if drift has also taken place, and it becomes necessary to take one continent (Africa) as a base. Wegener attempted to resolve the confusion concerning the controversial question of whether polar wandering arises from displacement of the whole crust over its substratum or from an internal shift of the axis of rotation. Some geophysicists reject the phenomenon because they assume that the equatorial bulge of the oblate Earth must remain unaltered in position. Wegener's belief that it can indeed move is put to empirical test by the marine 'transgression cycle'. The sea will follow at once if the equatorial bulge is reoriented but the solid earth will lag behind. Therefore in the quadrant ahead of the migrating pole one should observe from the stratigraphical record evidence of increasing regression; the reverse should apply to the quadrant in the rear. Wegener tried to show that this indeed is true between the Devonian and Permian.

As for the mechanism of continental drift, Wegener was tentative. 'The Newton of drift theory has not yet appeared. . . . It is probable that a complete solution of the problem of the driving forces will still be a long time coming. . . .' Nevertheless he could not resist a little speculation. Two possible components were recognized.

The *Pohlflucht* force was invoked to account for movement of continents towards the equator. This is a differential gravitational force resulting from the fact that the Earth is an oblate spheroid, being relatively flattened at the poles. It was first proposed early this century by Eötvös and its reality was widely accepted, although its power to move continents was very much in dispute. Another force must be invoked for the presumed westward drift of the continents. Wegener favoured some kind of tidal force and argues that retardation of the Earth by tidal friction must chiefly affect the outermost layers, which therefore should slide as a whole or as detached continental fragments over the interior. He noted another possibility, namely the existence of deviations in the Earth's shape from that of a smooth spheroid, as deduced from measurements of the gravity field. The existence of bulges locally should create a force tending to move crust laterally from topographically high to low regions. Such a force should be stronger than the *Pohlflucht* force and should also act on the sima.

In view of the general hostility with which these ideas, albeit tentative, were received, Wegener might have been tactically wiser to have ignored the problem of mechanism in the almost total absence of relevant evidence, and to have concentrated on marshalling the impressive array of data which seemed to favour drift, and criticizing the conventional stabilist views. On the other hand it could be argued that some people would have been converted only if they could have brought themselves to believe in some sort of working model of the Earth, however crude, which allowed the lateral displacement of continents to occur. Wegener can therefore hardly be criticized for trying.

3. The response to Wegener before the Second World War

> The mind likes a strange idea as little as the body likes a strange protein and resists it with similar energy. It would not perhaps be too fanciful to say that a new idea is the most quickly acting antigen known to science. If we watch ourselves honestly we shall often find that we have begun to argue against a new idea even before it has been completely stated.
>
> WILFRED TROTTER

WEGENER's early publications on continental drift do not appear to have attracted widespread attention. The interest of British and American geologists in the matter seems to have begun in 1922, when Philip Lake published a fair-minded but sceptical review of the second edition of *Die Entstehung der Kontinente und Ozeane*. Later that year, the British Association held a discussion on the hypothesis at the annual meeting in Hull. According to the published report by W. B. Wright, the discussion proved 'lively but inconclusive'. Scepticism was not unnaturally expressed, but the general reception accorded the new ideas seems to have been by no means unsympathetic. A more important symposium, organized by the American Association of Petroleum Geologists, was held a few years later in New York. The symposium publication contains articles by a number of the leading geologists of the day, who had by now had ample time to give a considered opinion. By and large the judgement was adverse to Wegener. Of the twelve contributors apart from Wegener and Taylor, only one, Van Waterschoot van der Gracht, could be described as strongly sympathetic. Nearly all the rest were hostile in various degrees, none more so than the American R. T. Chamberlin, the whole tone of whose article betrays a barely suppressed fury.

It would be both tedious and pointless to deal comprehensively with the many people who put forward criticism of the drift hypothesis in the pre-war period, since so much repetition is involved. In fact most of the important criticisms can be found in the A.A.P.G. symposium volume published in 1928. I shall instead outline the principal criticisms and mention a few of the more significant names associated with them.

The supposed jigsaw fit of the Atlantic continents was challenged on the grounds of inaccuracy. It was thought particularly inappropriate that only shorelines had been matched. In view of the abundant evidence that many of the world's shorelines have frequently been subject to epeirogenic (i.e. verticle tectonic) movements involving either uplift or subsidence, we would naturally expect a considerable amount of distortion over the time involved, of the order of tens or even more than a hundred million years. A close match of the shorelines in such circumstances could hardly be expected; it would in fact be a remarkable coincidence. Wegener had indeed made no serious attempt to examine the detailed shapes of shorelines and his only relevant portrayal of the continents in their changing positions is extremely crude (Fig. 3).

The closeness and significance of the geological similarities on the opposing sides of the Atlantic were also questioned. The American petrologist H. S. Washington had attracted widespread interest by an article pointing out that the igneous rocks mentioned by Wegener were not in fact all that similar. Schuchert and others likewise doubted how closely similar the stratigraphical sequences were, and wondered anyway whether similarity necessarily implied contiguity. Schuchert also scoffed at the evidence of Palaeozoic transgressions and regressions cited by Wegener in support of his notion of polar wandering. The actual pattern was far more complicated.

Wegener's references to glacial evidence came under attack from the leading Ice Age authorities, A. P. Coleman and R. T. Chamberlin. Concerning the supposed Permo–Carboniferous ice-sheet of Gondwanaland, Coleman made the point that much of the area in question would have been far from the ocean on Wegener's reconstruction, and hence out of reach of moisture-laden winds. Such a region would be more like central Asia, which never receives heavy snowfalls despite having the coldest temperatures recorded on Earth. Chamberlin thought that the matching of moraines to prove continental linkage in the Quaternary was positively ludicrous.

Charles Schuchert, the stratigrapher and palaeobiogeographer from Yale University, was one of the few to address himself to the faunal and floral similarities between continents. Certainly they existed; in fact it would be ridiculous to deny them. They were perfectly explicable, however, given only a few land bridges, such as across the Arctic, the South Atlantic, and the Indian Ocean, which subsided after the Mesozoic. Furthermore, the different faunas were similar but not identical, which one would expect if they formerly lived side by side.

The evidence of geodetic measurements was generally dismissed as inconclusive, and indeed the latest, post-war data seem to support this negative assessment. We may note parenthetically that, despite Wegener's evident fondness for this type of evidence, a failure to demonstrate conclusively that drift has taken place in the last few years or decades has little relevance to the general hypothesis.

Judging by the repeated comments that one reads in the literature or hears in conversation with geological or geophysical elder statesmen, the greatest stumbling-block to acceptance of Wegener's hypothesis involved the nature of the underlying mechanism.

Why, it was often asked, did 'Pangaea' remain a coherent super-continent for so long in the Earth's history and then break up so dramatically within a mere few tens of millions of years? Wegener and Taylor confined themselves very largely to the formation of the Tertiary fold-mountain belts. How were the older Caledonian and Armorican belts supposed to form?

How was it that the continental sial could move by displacing the sima and yet crumple under compression at its leading edge to form the Cordilleran ranges? Did this not imply significant backthrust from the supposedly weaker sima? How indeed was one to resolve this apparent contradiction—was it by distinguishing between *strength* and *rigidity* as Van der Gracht had done in the A.A.P.G. symposium?

How could the movement of the Americas from the Old World be described merely in terms of westerly drift, when the position of the Mid-Atlantic Ridge could be held to argue for an *eastward* movement of Africa, as Taylor had recognized?

Among the most formidable of Wegener's opponents was the Cambridge geophysicist Harold Jeffreys, the first edition of whose great treatise *The Earth* appeared in 1924, and the second in 1929. Somewhat scornful attention is paid to continental drift in both volumes, as also in his popular short book *Earthquakes and mountains*.

The essence of Jeffreys's argument is that the Earth has far too much strength to be deformed by the proposed tidal and *Pohlflucht* forces. The crust is evidently strong enough to hold up high mountain ranges and hold down ocean deeps. Tidal friction and the differences between values of gravity at the tops and bottoms of continents are capable only of producing stresses of the order of 10^{-5} dynes per square centimetre, whereas merely to elevate the Rocky Mountains the necessary value is close to 10^9. If, indeed, tidal forces were strong enough to allow westerly drift of the continents, the Earth's rotation would stop within a year. It was, Jeffreys said, very dangerous to make the assumption that the Earth could be deformed indefinitely by small forces provided that they acted persistently over long periods of geological time. Likewise, polar wandering was dismissed by Jeffreys and other geophysicists as being physically impossible.

Jeffreys noted that the widespread assumption of perfect isostatic compensation at the surface, which appeared to suggest an Earth of finite viscosity, so that rock could be distorted by small but long-acting shearing stresses, was not borne out by data then recently acquired. He regarded as especially absurd the notion that such a condition could extend up to the ocean floor. If it did, this would be perfectly flat, thus contradicting all the submarine soundings ever made. He also took strong exception to Wegener's invoking the *Pohlflucht* force to account for westward drift of continents, that is in a direction at right-angles to the imposed force. He pointed out that while a meteorologist might reasonably be expected to be impressed by the evidence of winds, whose movements are indeed dominantly parallel to the equator because of deflection by the Earth's rotation, the viscosity of rocks is far too high for this effect to be significant. On this point Jeffreys seems less than fair to Wegener, who invoked quite different forces for longitudinal and latitudinal movements of continents. Evidence of lateral, compressive movement in mountain belts was to Jeffreys a mere red herring: the movements were only of the order of 50 km or so. As for the geological and biological evidence for drift, Jeffreys contemptuously dismissed this in a few sentences.

A contracting Earth was still widely favoured, although Kelvin's model had of course been discredited by the discovery of radioactivity. Jeffreys preferred an evolutionary model whereby the Earth is supposed to have condensed as a liquid from a gas cloud, which subsequently cooled by convection after gravitational separation had rapidly formed the crust and core. Others preferred the 'cold accretion' or planetesimal hypothesis of T. C. Chamberlin, which involved compaction under gravity, allied with a rise in temperature, of small meteoritic bodies.

Some critics could not resist questioning Wegener's respectability as a scientist. Consider the following three quotations.

Whatever Wegener's own attitude may have been originally, in his book he is not seeking truth; he is advocating a cause, and is blind to every fact and argument that tells against it. Much of his evidence is superficial. Nevertheless he is a skilful advocate and presents an interesting case (P. Lake).

[Wegener's method] in my opinion, is not scientific, but takes the familiar course of an initial idea, a selective search through the literature for corroborative evidence, ignoring most of the facts that are opposed to the idea, and ending in a state of auto-intoxication in which the subjective idea comes to be considered as an objective fact (E. W. Berry).

Wegener's hypothesis in general is of the foot-loose type, in that it takes considerable liberty with our globe, and is less bound by restrictions or tied down by awkward, ugly facts than most of its rival theories. Its appeal seems to lie in the fact that it plays a game in which there are few restrictive rules and no sharply drawn code of conduct (R. T. Chamberlin).

Wegener made what seems a rather half-hearted attempt to reply to his critics in the last edition of his book. As far as one can judge from the account of his brother, Kurt, he felt rather overwhelmed by the flood of literature and did not feel he could cope with the drastic revision that some might have recommended. Thus the final edition is closely modelled on the earlier work but with some interesting additions to the supporting evidence. This evidence did not introduce anything strikingly new, and was hardly of the type to convince those sceptics who had been unconvinced by the data and arguments presented earlier. Since he died so soon afterwards, the rather touching opening statement of his book, gently rebuking the partiality of his critics, could appropriately serve as his epitaph: 'Scientists still do not appear to understand sufficiently that all earth sciences must contribute evidence towards unveiling the state of our planet in earlier times, and that the truth of the matter can only be reached by combining all this evidence.'

Though the consensus was quite evidently against him, Wegener had some formidable allies. The distinguished Harvard geologist R. A. Daly fully accepted that drift had occurred though he had serious reservations about the proposed mechanism. He suggested an alternative whereby the continents slid laterally under the influence of gravity owing to bulging of the polar and equatorial regions with a depression in between. Quite how such an unstable deformation was initiated was not explained. The two great synthesists of Alpine structures, E. Argand and R. Staub, wrote major articles in which lateral movement of continents played a significant role. Leading Dutch geologists working in the East Indies found that Wegener's ideas made more sense than those of the 'stabilists' in accounting for the complexities of structure. The outstanding British structural geologist E. B. Bailey was favourably disposed toward the hypothesis because it appeared to account so well for the similarities and crossing of the Caledonian and Armorican (or Hercynian) fold belts on the two sides of the North Atlantic.

Somewhat less formidable were a number of biogeographers who enthusiastically seized upon drift as an explanation of the distribution of their favourite organisms. They were handicapped by the inadequacy or total lack of a fossil record for the groups in question, and by their blithe ignorance or disregard of many basic facts of Earth science.

To my mind, however, quite the most significant pre-war supporters of Wegener were Arthur Holmes and Alexander du Toit.

Holmes, whom many regard as the greatest British geologist of this century, had established an absolute time-scale based on the constancy of rates of decay of radioactive elements in rocks. The thermal effects of radioactivity played a prominent role in a fascinating prophetic paper published in the *Transactions of the Geological Society of Glasgow* in 1929. In this subject he had been anticipated, and influenced, by J. Joly, who a few years previously had published a book entitled *Surface history of the Earth,* which dealt at length with the generation of radio-active heat, though from a standpoint hostile to continental drift.

While prepared to accept the evidence for continental drift (Fig. 9)

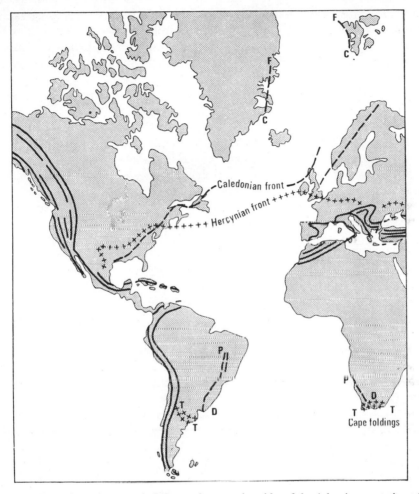

FIG. 9. The Palaeozoic orogenic belts on the opposing sides of the Atlantic, approximately as portrayed by Holmes in 1929.

Holmes was, like most others, dissatisfied with the proposed mechanisms of movement. He proposed a model of the Earth with upper, inter-mediate, and lower layers composed respectively of granitic, dioritic, and peridotitic (plus eclogitic) rocks. The *crust* consisted of the upper and intermediate layers together with the higher, crystalline part of the lower layer, and the *substratum* of the lower, glassy or thermally 'fluid' part of the lower layer. The sub-sedimentary basement of the ocean floor was composed of gabbro or amphibolite (a metamorphosed chemical equivalent of gabbro).

It had now become evident that ordinary volcanic activity as observed was insufficient to discharge the amount of heat from radio-activity in the substratum that was estimated to rise to the Earth's surface. The data made more sense if one invoked convection currents in the substratum which, if rising beneath continents, could cause continental drift.

In the simple case of a layer of uniformly heated viscous liquid with rigid conducting surfaces above and below, a condition of convective stability will exist until a certain critical temperature gradient is reached. This temperature gradient will depend on the compressibility, conductivity, and viscosity of the fluid. As the critical gradient is exceeded some form of convective circulation will begin. Local centres will develop at which currents rise. At the top of the layer these will spread, interfere, and turn down to produce a system of irregular polygonal prisms. In the Earth the critical gradient is about 3°/km, and calculations suggest that the viscosity of the substratum is not too high for convection to occur.

Holmes went on to say that as at high temperatures the strength of materials rapidly diminishes, the substratum should be devoid of strength, contrary to what Jeffreys had argued. He thought that the secular variation of the Earth's magnetic field supported this inference (a point which would be disputed today), as did regional isostatic compensation within a depth of 60 km.

Because of the Earth's rotation, upward-moving currents would be deflected westwards and downward-moving currents in the opposite direction. 'Monsoon-like' currents might be expected to occur owing to the distribution of continental blocks and ocean floors. The radio-activity of continental rocks is higher (because uranium and thorium are concentrated in granites) and the temperature beneath them should be higher than under the ocean. Currents would therefore rise under the continents and spread in all directions towards the peripheral regions. Downward-moving currents would be strongest beyond the edges of the continents, where weaker currents from the oceanic regions would be encountered (Fig. 10).

Above the places where the currents rise and diverge a stretched region would arise in the continental crust and eventually fragments would be torn asunder, leaving a disruptive basin to subside as new ocean, into which much excess heat would be discharged. At the con-tinental margins, the crust would be in compression and the amphi-bolite layer thickened. At high pressure and temperature this might be

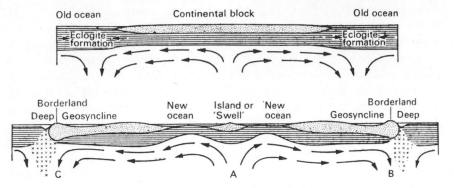

FIG. 10. Holmes's interpretation of continental drift. After Holmes 1929, Figs. 2 and 3.

expected to change into eclogite, which would result in a change in density from 2·9 or 3·0 to 3·4 or more. Blocks of eclogite would consequently sink into the substratum by a process known as stoping, and this would tend to speed up the downward-moving cooling current. In this way one could account for oceanic deeps such as those bordering the Pacific.

The upper, sialic, layer would also be thickened by differential flowage of its levels towards the obstructing ocean floor. This crustal thickening at the leading edge would lead to mountain-building. The mountain roots, being light and therefore unable to sink, would begin to fuse and give rise to the andesites and associated volcanic rocks such as those found so commonly in the 'Fiery Ring' of the Pacific borderlands. This mechanism for mountain-building associated with continental drift avoids the difficulty pointed out by critics of Wegener's ideas, with the sima being weaker than the advancing sial but nevertheless able to exert a backthrust. By Holmes's proposed mechanism, mountain-building would inevitably occur provided that the horizontal component of rock flowage were greater from behind than in front.

The velocity of subcrustal currents is estimated to be about 5 cm per year, probably when the excess of temperature above that corresponding to the critical gradient is 10°C. On this conservative estimate, the minimum period to produce, say, the Atlantic would be about 100 million years.

Holmes went on to outline a number of other geological consequences. He pointed out that the mechanism provides a ready explanation for geosynclines, the term given to large, elongate basins of sedimentation occurring characteristically at continental margins, frequently associated with volcanicity and converted subsequently into mountain ranges. Continental faulted rift valleys such as the great East African example stretching from Malawi to the Red Sea are also explicable in terms of subcrustal convection, though Holmes preferred a model whereby such rifts are formed under compression rather than tension, as was widely assumed. A final point of interest concerns the world-wide changes of sea-level inferred from the stratigraphical record. Behind the moving continents the material of the new and growing ocean floor is at first hot and expanded. As it rises in consequence sea-water is proportionately displaced on to the continents. As drifting slows down and the material cools a marine regression should follow. Holmes cited the major mid-Cretaceous transgression, followed by a late Cretaceous regression contemporary with a phase of mountain-building and followed in the Eocene by another large transgression.

In addition to this important paper, Holmes played a considerable role in making sure that the case of continental drift obtained a fair hearing both by means of review articles and his celebrated elementary textbook *Principles of physical geology.*

Du Toit was a quite different sort of geologist and personality. He had an impressive knowledge of the geology of his native country, South Africa, and had been much struck by the remarkably close resemblance of the Palaeozoic and Mesozoic geology of South Africa with eastern South America. A lengthy article of his on this subject in 1927 had caught the attention of Wegener, who enthusiastically referred to it in the last edition of his book. Du Toit became a leading disciple of Wegener and propounded drift theory indefatigably until his death. He is best known for his book *Our wandering continents*, published in 1937.

Du Toit set out initially, as did Wegener, to point out the inadequacy of 'orthodox geology' to account for a variety of major phenomena. Geosynclines and rift valleys have alternatively been ascribed to tension and compression; fold ranges to Earth contraction, isostatic adjustment, and igneous intrusion. Some have regarded the crust as weak, others as strong; some wanted the continents fixed in position, others allowed some limited movement. Wide or narrow transoceanic land bridges were allowed by many, despite geophysical objections. In contrast to all this, drift theory can explain in coherent fashion a wide range of geological phenomena without becoming involved in basic contradiction and confusion.

The rest of the book is devoted to improving on Wegener by amassing more supporting evidence and correcting what du Toit acknowledged as errors. Thus the argument of Quaternary connection between North America and Europe, based on terminal moraines, is abandoned, and the continents are fitted at the edge of the shelves rather than at the shorelines. The frequently expressed objection to Wegener, that his hypothesis failed to account for the formation of the pre-Tertiary

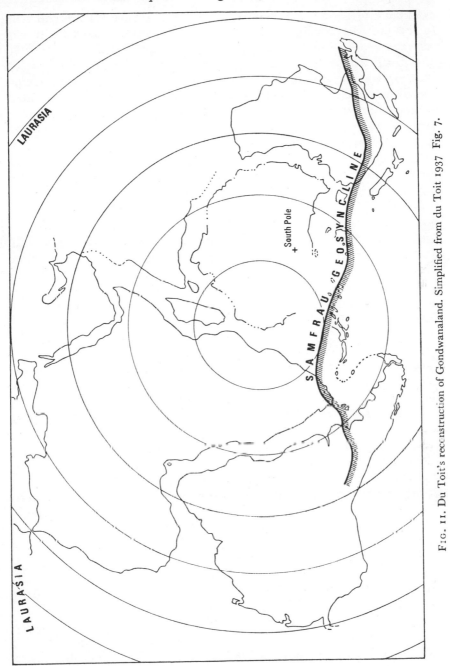

mountain belts, is countered by arguing that the Caledonian and Hercynian (Armorican) orogenies also involved continental drift of a sort.

A distinctive feature of du Toit's views is the separation of Wegener's 'Pangaea' into northern and southern supercontinents (Laurasia and Gondwanaland respectively) with somewhat independent histories, separated at least since the late Palaeozoic by an extensive seaway, 'Tethys', which was not broken up until the Tertiary as Africa and India drove northwards towards Eurasia.

Gondwanaland is reconstructed with far more accuracy than Wegener ever attempted (Fig. 11). On the basis of geological comparisons with the mainland, Madagascar is moved northwards from its present position with respect to Africa, to lie opposite Tanzania and Kenya. There is evidence, from the geology of fold mountains in Argentina, the Cape Province of South Africa, and eastern Australia, of a Palaeozoic geosyncline that du Toit called the Samfrau geosyncline (*South America*, South *Africa*, *Au*stralia) and the line of the different fold belts is used to guide the reconstruction. The numerous stratigraphical correspondences with Gondwanaland are discussed in considerable detail and presented in tabular form (Fig. 12).

The complete break-up of Gondwanaland was not achieved until Cretaceous and Tertiary times, but 'the bonds were relaxing' in the Jurassic. This is signified by a shallow marine incursion into the hitherto terrestrial region of East Africa and Madagascar, which probably signifies crustal subsidence under tension without true fracture. Persistence of terrestrial faunal connections until the Cretaceous suggests that sedimentation might periodically have filled up the gulf, or the sea floor might locally have bulged up to form temporary land bridges. The opening of the South Atlantic began in the north during the early Cretaceous and was completed before the close of the period, as shown by the oldest marine deposits on the eastern and western coastlands. India commenced its north-eastwards drift in the early Cretaceous. The late Cretaceous–early Tertiary basalts of the Deccan Plateau, covering 500 000 km^2 of western India, are clearly associated with drifting. Australia was still linked to Antarctica but not Africa in the early Cretaceous. It broke away from Antarctica at some time during the late Cretaceous and early Tertiary, with New Zealand drifting away from it in turn during the Tertiary.

With regard to Laurasia, the timing of the Atlantic opening paralleled movements in Gondwanaland. Initial crustal sagging in the Jurassic is evidenced by marine transgression, but major rupture and drifting did not take place until the Cretaceous and Tertiary. The late Jurassic Nevadan orogeny of California might be held to signify the initiation of westward movement of North America, with the end-Mesozoic Laramide orogeny of the Rockies marking a further phase of movement. During the Eocene the Iberian Peninsula was rotated anticlockwise, opening up the Bay of Biscay and causing compressional movements (folding and thrusting) in the Pyrenees. The Tertiary mountain belts of the world are grouped into two major groups. The Alpine–Himalayan ranges are clearly the result of the impingement of parts of the old Gondwanaland on Laurasia.

As for the mechanism, du Toit was not prepared to say much, but he expressed support for an interpretation along the lines that Holmes had proposed.

Du Toit's valuable work is spoiled to some extent by an unfortunate style. Like many disciples, he quite outdid his master in proselytizing zeal and took evident delight in belabouring the champions of orthodoxy. All this is very well, but the occasional use of a rapier rather than a cudgel might have been more effective. In his enthusiasm he tended to spurn any attempt to disentangle description from interpretation, which makes for exasperating reading. One sometimes gets the impression that all facts are grist to a particular mill and nothing is allowed to impede the flow of grand ideas. He tends to be prolix rather than laconic, and cannot resist dramatizing. Stratigraphical resemblances are 'astounding', the distribution of Carboniferous glacial deposits is 'extraordinary', terrestrial deposits are 'enormous' in extent, and different land-masses have 'wonderfully similar histories'.

The following passage from page 3 of his book gives a good example of his prose style. If the reader accepts continental drift, 'He will have to leave behind him—perhaps reluctantly—the dumbfounding spectacle of the present continental masses, firmly anchored to a plastic foundation yet remaining fixed in space; set thousands of kilometres apart, it may be, yet behaving in almost identical fashion from epoch to epoch and stage to stage like soldiers at drill; widely stretched in some quarters at various times and astoundingly compressed in others, yet retaining their general shapes, positions and orientations; remote from one another throughout history, yet showing in their fossil remains common or allied forms of terrestrial life; possessed during certain epochs of climates that may have ranged from glacial to torrid or pluvial to arid, though contrary to meteorological principles when their existing geographic positions are considered—to mention but a few such paradoxes!'

Splendid stuff undoubtedly, but this is the colourful language of a pamphleteer. Many a scientist reading *Our wandering continents*, accustomed to a literary tradition of sober, unemotive expression in the third person and passive voice, must have reacted with the same

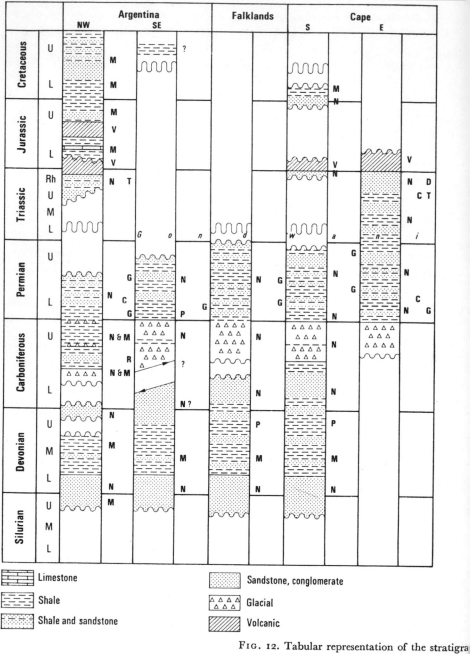

FIG. 12. Tabular representation of the stratigra

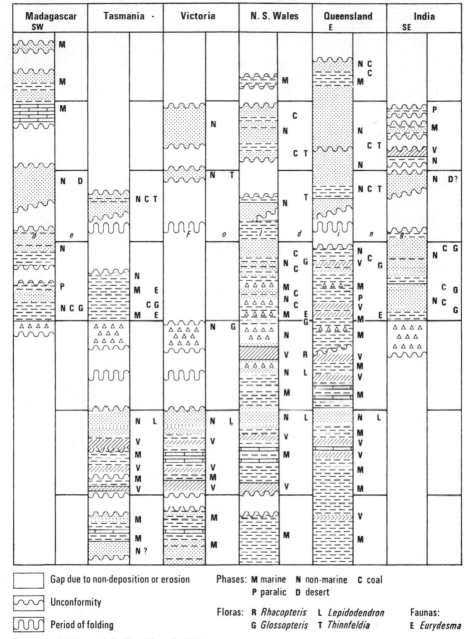

the southern continents and India. After du Toit 1937.

distaste as a philosopher of the Augustan age to the utterings of a religious enthusiast.

Despite his shortcomings du Toit unquestionably made a substantial contribution to the drift hypothesis, partly by eliminating some of Wegener's errors and partly by integrating a vast amount of evidence, much of it new, into a plausible story whereby a wide array of disparate facts was given a coherent, simple interpretation. Holmes, for his part, had at last outlined a plausible mechanism, contestible though it certainly was in places. His ideas might have been wrong but at least they were intellectually respectable, and warranted serious and careful consideration. One might therefore have expected, after the initial sceptical stir provoked by Wegener's hypothesis, that groups of scientists would have been sufficiently stimulated to put it to test in various ways.

This did not happen in the pre-war or even the immediately post-war period. Instead, continental drift was, by most scientists, especially in North America, either rejected outright as a farrago of nonsense or at least treated with considerable scepticism. Perhaps Holmes suffered from publishing in such an obscure journal, while du Toit's exuberant style laid him open to the charge of being a crank and hence easily dismissible. No doubt some of the small band of drift enthusiasts were cranky but one normally expects to judge an hypothesis by its best exponents. At any rate, the great majority of geologists and geophysicists were content to put large ideas such as continental drift to the back of their minds and concentrate upon their various specialties. It required an explosive outburst of new data and ideas from a much-enlarged scientific community in the 1950s and 1960s to point up the previous few decades as by and large a rather stagnant phase in the history of Earth science.

4. Post-war developments in new research fields

ONE IMPORTANT reason why the pre-war debate had proved so inconclusive was that we were in almost total ignorance about what underlay the oceans, which together with inland seas cover no less than 70 per cent of the surface area of our planet. Our knowledge was to be vastly increased from the 1950s onwards and many of our ideas subsequently transformed primarily as a result of the new oceanographic work, but the first major disturbance to the still-conventional world view of fixed continents came from another comparatively young research field, rock magnetism.

Rock magnetism, or *palaeomagnetism*, had undergone a slow development for several decades and received an impetus in the late 1940s and early 1950s from the desire of scientists to investigate the ancient geomagnetic field in the hope that it might throw some light on that of the present day. New, highly sensitive magnetometers were designed and statistical techniques developed for analysis of data.

The magnetization of rocks is in effect a fossil permanent magnetism or *natural remanent magnetism* (n.r.m. for short) which can serve as a kind of fossil compass to determine the direction of the ancient or palaeomagnetic field. The research method is based on the assumption that the mean geomagnetic field is that of an axial dipole situated at the Earth's centre. Oriented rock samples of certain types yield information on the field direction (giving information about the direction of the poles) and latitude, if the angle of declination D and the angle of inclination I are determined by a sensitive magnetometer. The angle of latitude L (see Fig. 13) is determinable from the simple equation

$$\tan I = 2 \tan L.$$

Only a few types of rocks can be used. Basaltic lavas are fairly iron-rich and acquire their magnetization from the geomagnetic field as they cool through the Curie points of their iron–titanium oxide minerals upon crystallization after they have erupted at the Earth's surface.

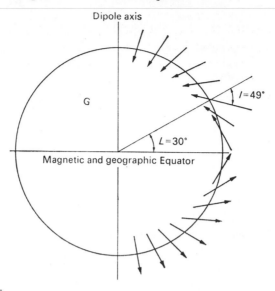

FIG. 13. Inclination (*I*) variation with latitude (*L*) of the geomagnetic field with magnetic and geographic poles coinciding. Arrows represent lines of force. Based on A. G. Smith 1971, Fig. 15.9.

Certain sedimentary rocks such as red sandstones contain enough iron oxide to acquire a measurable magnetization. The way in which such sediments become magnetized remains somewhat obscure. To some extent grains of iron oxide minerals may behave as minute magnets during sedimentation and align themselves in the field. Magnetization may also be acquired subsequent to deposition, however, during the process of chemical and physical alteration known as diagenesis. In determining *D* and *I* it is of course necessary to orient the rock sample as they would have been when they were formed; in other words the effects of subsequent tectonic dip disturbance must be eliminated.

The fundamental assumption concerning the axial dipole nature of the field is justifiable on both theoretical and empirical grounds. It is consistent with the widely accepted theory of the geomagnetic field associated with the names of Sir Edward Bullard and W. M. Elsasser, which argues that the fluid, electrically conducting outer core of the Earth acts as a self-exciting dynamo. Although at the present time the magnetic axis makes an angle of some $11\frac{1}{2}°$ with the geographic (rotation) axis, it is virtually coincident when averaged over periods of several thousand years. Furthermore, palaeomagnetic determinations on lava samples going back 10 million years or so into the Upper Tertiary indicate a position for the geographic North Pole that cannot be distinguished statistically from that of today.

By the middle 1950s some intriguing results were being obtained by research groups at Cambridge and Imperial College, London. The Cambridge group, notably S. K. Runcorn, K. M. Creer, and E. Irving, obtained samples of European rocks of widely different ages and were able to demonstrate a steady change with time, prior to the Upper Tertiary, of the position of the North Pole. From a position near Hawaii in the Pre-Cambrian it had slowly migrated north-westward to between Japan and Kamchatka by about the end of the Palaeozoic and thence via eastern Siberia to its present position (Fig. 14).

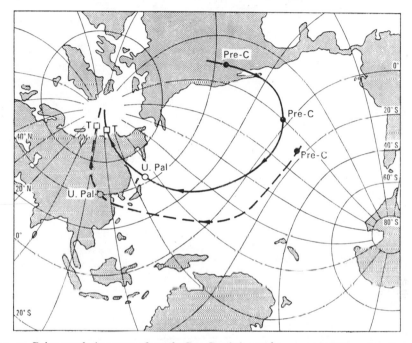

FIG. 14. Polar wandering curves from the Pre-Cambrian to the present in Europe (continuous line) and North America (broken line). Simplified from Runcorn 1962, Fig. 19.

As the differences compared with the present were not random but systematic, with the angle of polar difference increasing with time, a migration either of the poles or of the continents was clearly suggested. An independent check on the interpretation of change of latitude can be made using geological evidence of palaeoclimates. Thus the presence of thick coals and coral-bearing limestones in the Carboniferous and of evaporites and desert sandstones in the Permian of Europe are consistent with the palaeomagnetists' estimate of latitudes in those periods of between about 20°N and 20°S. The presence of large reptiles and reef-building corals in European Jurassic deposits is likewise consistent with an estimated latitude 10° to 20° less to the north than today. A third example of agreement is Australia. The palaeomagnetists' estimate of high palaeolatitude in the late Palaeozoic is supported by the existence of glacial deposits, and isotopic palaeotemperature data from Tertiary fossils support the suggested decrease of palaeolatitude from the Palaeocene to the Miocene (Fig. 15).

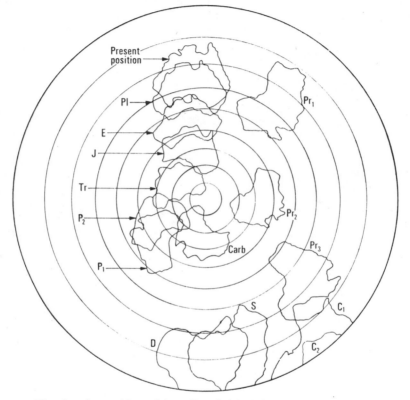

FIG. 15. The changing positions of Australia relative to the South Pole since the Pre-Cambrian. Pr, Pre-Cambrian; C, Cambrian; S, Silurian; D, Devonian; Carb., Carboniferous; P, Permian; Tr, Triassic; J, Jurassic; E, Eocene; Pl, Pleistocene. Adapted from Runcorn 1962, Fig. 22.

Now the recognition of apparent polar wandering was a surprising result because, as we have seen, conventional geophysical theory had firmly favoured the fixity of pole positions. Were the data therefore unreliable, or was the theory in error? Fortunately an English astronomer, T. Gold, quickly came to the rescue in 1955. An analysis of minor rotational perturbations other than precession was only explicable if there were *no* permanent stiffness in the geoidal shape of the Earth, as had previously been claimed, and Gold was able to trace the error back to Sir George Darwin. Since plastic flow is possible, even a slight redistribution of mass will tend to cause the poles to wander, despite the stabilizing effect of the equatorial bulge. For instance, the raising of a continent the size of South America by 30 m in a few million years, which is not geologically unreasonable, should cause the poles to rotate by up to 90°. Since this evidently has not happened, the problem becomes one of finding a suitable damping agent, and Gold thought that the polar ice-caps might fulfil this function.

Some cynical geological observers were not slow to point the moral that our knowledge of the physics of the Earth's interior is so limited that the detection of a single error can lead to a dramatic reversal of interpretation.

An even greater surprise was to follow from the palaeomagnetic work shortly afterwards. Runcorn and Collinson followed up the analysis of European data by doing the same thing for North American rocks. Once more a regular polar wandering path was determined which was broadly similar to the European one, but a systematic difference was discernible, the American Pre-Cambrian and Palaeozoic path being displaced about 30° of longitude to the west (Fig. 14). After the Triassic the difference disappeared. It is to Runcorn's credit that, although like nearly all geophysicists he had been taught to believe in the unlikelihood, if not impossibility, of continental drift he quickly recognized that the anomaly could be made to disappear if the North Atlantic were closed by bringing North America adjacent to Europe, as Wegener and Taylor had proposed. The implication was that the continents had moved apart at some time between the Triassic and the present. From then onwards (1956) Runcorn became a leading exponent of the lateral migration of continents.

It was not long before a series of results were reported from rocks collected in the southern continents. Once again, a systematic change with time through the Palaeozoic, suggestive of polar wandering, was recognizable, and divergences of polar wandering paths for different continents could be eliminated if they were brought together as Gondwanaland.

This new work naturally stimulated great interest and, if it failed immediately to make large numbers of converts to continental drift, at least many Earth scientists became tolerant towards the hypothesis. By no means all the reported palaeomagnetic results which flooded the literature for the next few years were easy to interpret. Indeed, gross and apparently inexplicable anomalies appeared quite frequently, providing documentary material for the sceptics of the method.

Much of the trouble arose because of the disturbing effects of secondary magnetization which may be acquired long after formation of the rocks, or from weak and inadequate original magnetization. The more refined modern approach is to be, in the first place, more selective in the choice of sample material, and in the second place to get rid of the unstable, secondary components of magnetization. This can be done by subjecting the samples to alternating magnetic fields or heat treatment, known respectively as magnetic and thermal cleaning.

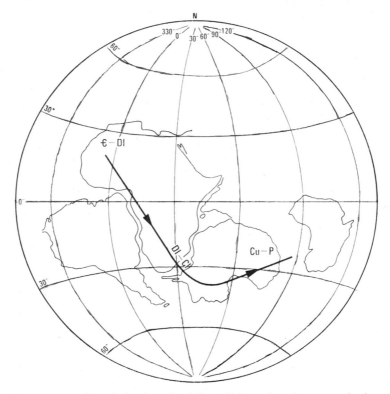

Fɪɢ. 16. Reconstruction of Gondwanaland from Palaeozoic palaeomagnetic data, with Africa placed in its present coordinates. The common polar wandering path is shown by the thick line. Simplified from McElhinny and Luck 1970, Fig. 1.

A recent comprehensive analysis of palaeomagnetic data from all the southern continents, by M. W. McElhinny and G. R. Luck, points to a reconstruction of Gondwanaland strikingly similar to that proposed originally by du Toit (Fig. 16). Also in support of du Toit, other data, analysed independently by McElhinny and J. C. Briden, indicate a Cretaceous break-up of the supercontinent. The difference in timing and magnitudes of episodic Palaeozoic polar shifts in Gondwanaland on the one hand and Laurasia on the other led McElhinny and Briden to suggest that the movements must have involved continental drift.

Furthermore, although one or perhaps two supercontinents existed in the Upper Palaeozoic, Laurasia was probably a series of separate fragments prior to this.

One of the more disturbing findings of the earlier palaeomagnetic work was that Laurasia and Gondwanaland, while being interpretable from the data as former supercontinents along the lines predicted by du Toit, overlapped slightly in position on the globe. This is geologically absurd, and led to the proposal that one land-mass had been shifted some 5000 km along the Tethyan zone with respect to the other by an enormous transcurrent fault or 'megashear'.

A new analysis of this problem by Briden and his co-workers rules out the Tethyan megashear or 'twist' as being geologically unreasonable and favours the Permian-Triassic reconstruction illustrated in Fig. 34. A plot of the best palaeomagnetic data of these periods now available leads to the suggestion that the geomagnetic field of that time approximated to that of a geocentric dipole, but that second-order departures from this cannot be dismissed as errors in the data and are most reasonably interpreted as multipole components of the field.

Quite a distinct field of palaeomagnetism concerns itself with short-period reversals of polarity, first observed when successive members of late Tertiary or Quaternary lava sequences were shown to have n.r.m. directions 180° apart. Initially it was by no means clear whether these changes signified true reversals of the geomagnetic field or self-reversal magnetization related to the iron–titanium oxide mineralogy of the rocks. The fact that self-reversal is extremely hard to produce experimentally, and that the direction of magnetization is the same in widely different but adjacent rock-types, such as basaltic lavas and sills and their baked sedimentary country-rocks, argues in favour of the former interpretation.

The most convincing evidence in favour of geomagnetic reversals comes, however, from dating the lavas by measuring the amount of radioactive decay of potassium-40 to argon-40. If it can be shown that rocks of the same age in widely separated parts of the world are magnetized in the same direction then it becomes very hard to dismiss the data as being due to random, local effects. This in fact was first done convincingly in 1963 by two groups of palaeomagnetists. A. Cox and his co-workers agreed closely with I. McDougall and D. H. Tarling that the last field reversal had taken place one million years ago and the last previous reversal about one-and-a-half million years earlier.

One might well wonder what possible relevance short-term reversals of the geomagnetic field have to continental drift. The answer is a fascinating one but cannot be gone into until the modern oceanographic work has been reviewed.

The great post-war advances in geological and geophysical oceanography have resulted from the deployment of large resources and the development of new techniques. Not surprisingly in the circumstances, the field has been even more than normally dominated by the Americans, notably those working at the Scripps Institute of Oceanography in California and the Lamont–Doherty Geological Observatory of Columbia University, New York.

We may consider first the results of topographic and geological exploration in the 1950s. Newly developed echo-sounding devices have allowed much more refined mapping of sub-oceanic topography than had previously been possible. The oceans may be subdivided topographically into three major provinces: the *continental margins* (comprising the continental shelf, slope, and, locally, deep trenches); the *ocean-basin floor* (abyssal floor, ocean rises, and seamounts); and the *mid-oceanic ridges*, each accounting for about a third of the total area. Geological samples were obtained by dredging and using shallow coring devices which penetrated a few metres into soft sediment.

The form of the continental shelves and margins appeared to be determined partly by major fault structures and partly by the out-building of shallow-water sediments. The topography of the abyssal plains was normally very smooth and they appeared to be sediment-filled basins. Cores of the sediment revealed either pelagic deposits such as red clay, foraminiferal oozes, and radiolarian oozes, or an alternation of these with thin layers of sand, thought to have been carried to the deep sea from the shelf by turbidity currents. Seamounts rising from the ocean-basin floor consisted of basalt and were evidently volcanic in origin. Some of these, for instance in the west Pacific, are flat-topped. The occurrence on their tops of Cretaceous corals and other fossils clearly indicative of shallow water, now at depths as much as 1600 m, pointed to a history of subsidence comparable with that of associated atoll groups. Nowhere did dredging of exposed rock or coring of sediment reveal rocks older in age than Cretaceous.

The mid-oceanic ridge system is without question the most spectacular feature and compares topographically with high mountain ranges on land. Its topography is rugged and suggestive of fault-block fracturing, and the ridges seem to form a broad, continuous swell that B. C. Heezen and others claimed to trace for 35 000 miles through the oceans (Fig. 17). In fact the 'ridges' may consist rather of broad rises or swells over 1000 km wide, as in the case of the East Pacific Rise. Of especial interest is the axial region, which is correlated with a zone of shallow-focus earthquakes. Its relief in some areas suggests a central rift valley (Fig. 18).

Such a median valley was first discovered by the British oceano-

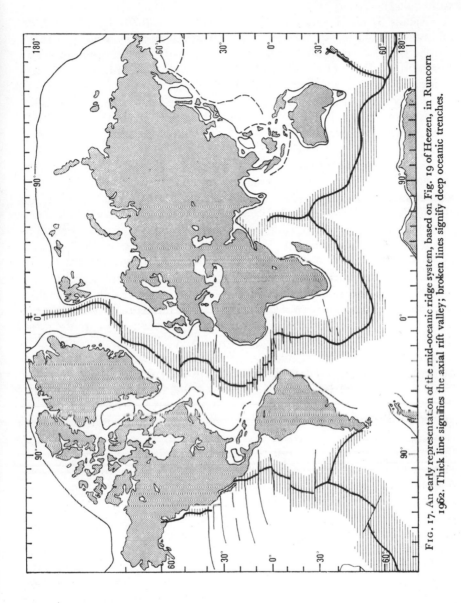

FIG. 17. An early representation of the mid-oceanic ridge system, based on Fig. 19 of Heezen, in Runcorn 1962. Thick line signifies the axial rift valley; broken lines signify deep oceanic trenches.

grapher J. C. Swallow on a cruise of the *Challenger* in 1953. It was subsequently traced into the western Indian Ocean ridge system by the Americans B. C. Heezen and M. Ewing, who postulated on grounds of broad similarity of topography and seismicity, as well as apparent geographic continuity, that the axial valley continued into the East African Rift via the Gulf of Aden. The mid-oceanic ridge valley tends to be about 25 to 50 km wide and 250 to 750 m deep, and is frequently flanked by steep scarp faces. Its popular interpretation in the late 1950s and early 1960s as a rift valley, indicative of crustal tension, was to be great significance in the understanding of the ocean floor.

Another branch of the ocean ridge system, the East Pacific Rise, enters the Gulf of California and appeared to pass under the high plateau and mountain country of the western United States, though as we shall see this is not the interpretation accepted today.

Dredge samples of exposed rock-surfaces on scarps in the mid-oceanic ridges have revealed a variety of fresh or metamorphosed basic igneous rocks, such as basalt, gabbro, and amphibolite.

The geophysicists have successfully studied the crustal structure by using gravity and seismic data in conjunction. The gravity-measuring technique was revolutionized during the International Geophysical Year by the development of gravimeters which could operate on ships during a large proportion of a voyage and thus made continuous profiling possible. The seismic techniques depend on the measurement of seismic energy propagated through sub-oceanic rocks by artificial explosions. The sound energy is received at varying distances from the explosion site by hydrophones or geophones and recorded on oscillograph tapes with an accurate time-scale. If the distance and time of travel are known the speed of propagation can be computed. With a line of receiving positions a relation can be established between distance and travel time and hence the number and thickness of subsurface layers determined. The speed of sound through a given layer gives an indication of the type of rock of which it is composed.

Two techniques must be distinguished: *seismic reflection* and *seismic refraction*. Interfaces of rock strata are often good reflectors, and so the reflection method is analogous to echo-sounding. The layer thickness is easily determined and continuous profiling gives a detailed structure. High sonic frequences are required to give a good resolution of sea floor and subsurface stratal topography. This technique is an excellent one for detailed geological investigations of sedimentary layering near the surface and its relationship to shallow basement rocks. High-frequency sound cannot penetrate deep into the subsurface, however, because of absorption, so for analysis of the crust as a whole seismic refraction

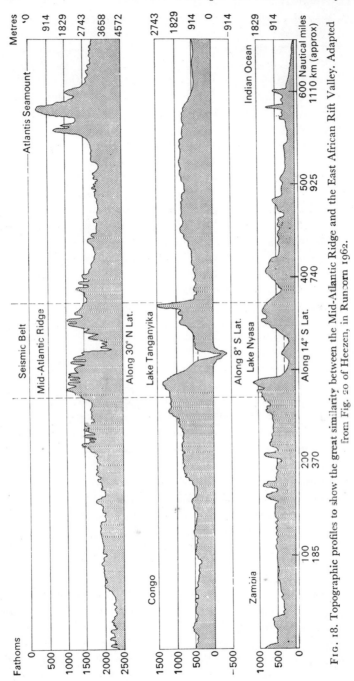

Fig. 18. Topographic profiles to show the great similarity between the Mid-Atlantic Ridge and the East African Rift Valley. Adapted from Fig. 20 of Heezen, in Runcorn 1962.

methods, utilizing low sonic frequencies, are used, preferably in conjunction with gravity determinations.

The most striking finding was that the depth to the base of the crust,
or Mohorovičić discontinuity (*Moho* for short), below which velocities
of seismic wave propagation increased suddenly to more than 8 km/s,
diminished sharply below the continental slope from an average value
of 35 km/s under the continents to 12 km/s below sea-level in the
oceans. Furthermore, except for a few small areas such as the Seychelles
Bank in the Indian Ocean, low-density 'sialic' rocks typical of the
continents were absent, so confirming what had long been suspected by
some on the basis of gravity observations and the analysis of earthquake
waves (Fig. 19).

Three crustal layers are usually distinguished, numbered 1 to 3 from
top to bottom. Layer 1 is typically about 1 km thick, with a transmission
velocity of about 2 km/s. It is attributed to sedimentary rocks. Layer 2
is about 1·7 km thick on average (5·1 km/s) and was considered to be
either consolidated dense sediment such as limestone or a modified
version of layer 3 (recent deep boreholes indicate the latter). Layer 3
averages nearly 5 km in thickness (6·7 km/s) and is widely considered
to consist of gabbro or basalt.

As for what used to be called the substratum beneath the crust and
is now universally called the mantle, this is widely thought to consist
under the oceans of the ultrabasic rock composed of olivine and
pyroxene known as peridotite. This is on the threefold evidence of the
transmission velocity of 8·1 km/s, of estimates of density, and of the fact
that peridotite upon partial melting yields a magma with the composition of basalt, the volcanic rock ubiquitous in the oceanic sector.
(Under the continents, at lower levels and hence higher lithostatic
pressure, another rock, eclogite, might also be present. This is similar in
chemical composition to basalt but contains higher-pressure phase
minerals of the garnet and pyroxene families.) Not everywhere under
the oceans is the uppermost mantle transmission velocity over 8 km/s.
Locally under the mid-oceanic ridges, a lens-shaped zone with velocities
between 7 and 8 km/s is recognized (Fig. 19).

A final point of interest about the mantle structure is the confirmation
of a world-embracing low-velocity layer at about 100 to 200 km below
the surface. This had first been predicted to occur beneath the continents by the seismologist B. Gutenberg, on the basis of attenuation of
earthquake shear waves, and the layer is consequently sometimes called
after him.

Two other geophysical techniques, involving the measurement of
heat flow and rock magnetization, have yielded results of great interest.
The rate of flow of heat through the ocean floor is determined as a

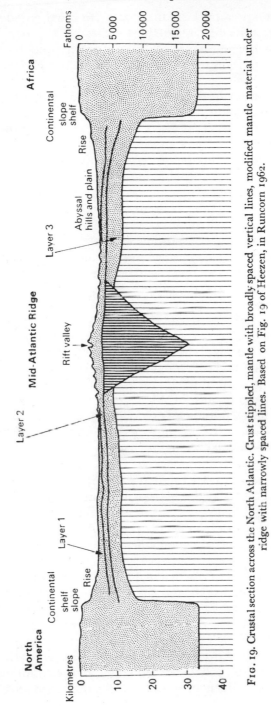

FIG. 19. Crustal section across the North Atlantic. Crust stippled, mantle with broadly spaced vertical lines, modified mantle material under ridge with narrowly spaced lines. Based on Fig. 19 of Heezen, in Runcorn 1962.

product of temperature gradient and thermal conductivity. The temperature gradient is measured by forcing a probe containing a recorder with a temperature-sensitive element into the ocean floor; thermal conductivity is measured on sediment samples collected with a coring tube.

Average values of heat flow through the floors of the ocean basins were found to be hardly distinguishable from, if not identical with, those of the continents. This was a most unexpected result, because the much higher radioactivity of continental rocks should have resulted in appreciably higher values being recorded than in the oceans. G. J. F. Macdonald suggested in explanation that the mantle under the oceans must be fundamentally different than that under the continents down to depths of several hundred kilometres at least, and that more radio-activity was present at depth in the oceanic mantle. This would lead to an equalization of surface heat-flow values.

Heat flow under the axis of the mid-oceanic ridge system was found to be significantly higher than elsewhere in the oceans, which was not surprising considering the evidence of volcanicity in various places such as Iceland, which lies astride the Mid-Atlantic Ridge and in fact marks a place where it rises to the surface.

To make total magnetic field measurements at sea, magnetometers with gyrostabilized platforms are towed behind ships or aircraft at a sufficient distance to avoid magnetic disturbance due to the engines or hull. Most, if not all, of the magnetism measured comes from the magnetite-bearing basalts of the ocean floor. A series of *magnetic anomalies* is recorded graphically on continuous traces. These anomalies represent small departures, measured in milligauss, from the mean values of total magnetic intensity measured in the direction of the geomagnetic field.

The technique was used to great effect in the late 1950s in the east Pacific by researchers at the Scripps Institute. They found a persistent north–south alignment of the anomalies, which on east–west traverses gave a characteristically oscillatory pattern of positive and negative values about the mean. The anomalies were tentatively held to delineate elongate bodies of magnetite-bearing rock, presumably basalt, aligned north–south, with patterns of magnetization sharply contrasted with their neighbours to the east and west. Quite what controlled the variation was not apparent. The researchers do not seem to have been unduly disturbed by this because they had become excited by something else.

The east Pacific is crossed by a number of major topographic features aligned east–west, such as the Mendocino Escarpment off California, which strongly suggested faults. In the Atlantic, comparable structures could be seen to displace the axial ridge (Fig. 17). It was

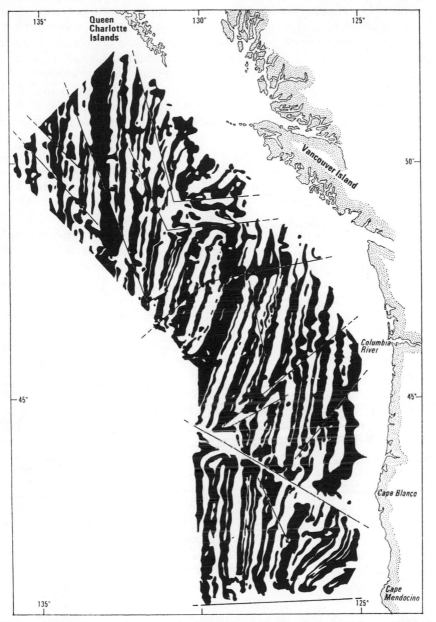

FIG. 20. Magnetic anomaly patterns in the north-east Pacific off the Canadian and United States coasts. Straight lines indicate faults displacing the anomaly pattern. After Vine 1966, Fig. 1.

found that the unique oscillatory pattern of given anomaly traces could be accurately matched across these presumed fracture zones, but only by invoking lateral shifts of ocean floor of the order of hundreds of kilometres (Fig. 20). This generally accepted finding carried with it a striking implication. The oceanic crust evidently behaved with a rigidity comparable to that of the continents, and blocks of crust could apparently be displaced large distances with respect to each other. Incidentally, none of the east–west oceanic faults seemed traceable into the North American continent.

This naturally leads us on to consider briefly some interesting tectonic work on the continents that was also being done in the 1950s. Faults in which the predominant movement of one block against the other is lateral rather than vertical are known variously as transcurrent, wrench, or strike-slip faults. Two types are distinguished. Imagine the rather alarming experience of standing astride such a fault when the ground starts to shift. If the right side moves towards you we call the fault *dextral* or right lateral, and *sinistral* if the reverse applies.

The Great Glen fault in Scotland is now generally considered to be a sinistral transcurrent fault, north-west Scotland having moved southwards relative to the rest of the country some hundred kilometres at some time since the mid-Palaeozoic. A far more spectacular example, however, is the San Andreas fault of California, which extends from north of San Francisco to east of Los Angeles. This fault is dextral, because the western sector of California can be proved to have been moving northwards within the last century at an average rate of 5 cm/year. Each minor phase of movement is, of course, recorded by an earthquake. By matching distinctive rock-types on each side of the fault, it has been confidently argued that the western sector has moved no less than 260 km since the Oligocene and perhaps as much as 560 km since the Jurassic. This implies an average rate of movement of almost 1 cm a year over a time-interval exceeding 100 million years.

Another well-known transcurrent fault where movement of hundreds of kilometres has been inferred is the Alpine fault of New Zealand. This is also dextral and so are nearly all the major transcurrent faults that roughly parallel the margins of the Pacific. Such a remarkable pattern actually led to the rather fanciful suggestion that the Pacific might be slowly rotating anticlockwise with respect to the adjacent continents.

The important point to grasp is that movement of continental and oceanic blocks on this scale and at this pace renders the notion of drift much more plausible, even though the actual motions involved were not those proposed by Wegener and du Toit.

It is also appropriate here to mention a lengthy paper published in

1958 by the Australian structural geologist S. W. Carey. Whereas European tectonicians at this time were becoming increasingly preoccupied with minutiae such as minor fold-structures, lineations, and even the orientation of crystal fragments seen under the microscope, Carey made a bold return to the approach of Suess and Argand in taking whole fold belts and other major tectonic features as his subject-matter. He introduced a number of new terms, some of which have proved useful. An *orocline* is an orogenic belt with a change in trend (e.g. from the Alps to the Apennines) which is interpretable as an impressed strain. A *sphenochasm* is a wedge-shaped sector of oceanic crust separated by continental blocks which originated by rotation or pulling apart of the latter. A *megashear* is a transcurrent fault of huge dimensions.

Carey's technique was in essence to straighten out the oroclines, close the sphenochasms, and reverse the megashears. The end-result was an Earth in which the continents were closed up in Wegenerian fashion in the late Palaeozoic but with a diameter only about three-quarters of the present value. The continents had, in his view, not drifted apart but dispersed as the Earth expanded at an evidently fast rate.

Carey's work was looked on askance by many more meticulous minds, and indeed some of his reconstructions seem quite insupportable, but there is no denying his stimulating influence, if only in encouraging structural geologists to turn their attention more to the wood than the trees. The ironic thing is that Carey's ideas were amenable in some instances to independent test. For this reason they were actually more, rather than less, scientific than the smaller-scale operations of some of his critics.

The tests in question are palaeomagnetic. The drastic Earth expansion proposed by Carey and a few others was tested by this means and found unacceptable, as indeed it seems to be on other grounds as well. On the other hand his proposal, which really can be traced back to Argand, that the Bay of Biscay is an oceanic sphenochasm opened up by the anticlockwise rotation of Spain away from France was supported by palaeomagnetic evidence. Likewise, palaeomagnetic work confirms Carey's view that both Italy and Corsica–Sardinia have rotated anticlockwise, thereby opening up the Tyrrhenian and Ligurian seas.

The new data from widely different fields of research increasingly made the Earth science community more amenable to the idea of continental drift, but it was by no means clear how all these diverse findings could be integrated into a coherent picture, and problems were arising faster than they were being solved.

By the start of the 1960s the time was ripe for a few bright ideas.

5. The spreading sea-floor hypothesis

The joy of discovery is certainly the liveliest that the mind of man can ever feel.

CLAUDE BERNARD

IN RETROSPECT it is not particularly surprising that the required flash of insight which was to launch a new era in Earth science should come from the late Harry Hess of Princeton University. In the course of a distinguished career his research interests had ranged more widely than most, from the mineralogy of pyroxenes and the origin of serpentine bodies in mountain belts to submerged sea-mounts and deep oceanic trenches. He was especially intrigued by the pre-war discovery of the Dutch geophysicist F. A. Vening Meinesz that the trenches occurring immediately oceanward of island arcs in south-east Asia are characterized by pronounced negative gravity anomalies.

It was indeed apparent to many that the circum-Pacific arc–trench system, marked by numerous volcanoes and powerful earthquakes suggestive of faults dipping to great depth away from the oceans, held the key to unravelling the mysteries of geosynclines and mountain-building. Hess shared Vening Meinesz's view that the strong departure from isostatic equilibrium of trenches which descend as much as 4 km below the oceanic abyssal plain was best accounted for by their being held down by some subcrustal force such as the descending limb of a convection current.

Another of Hess's notable scientific contributions, arising directly from his wartime exploits as a commander in the U.S. Navy, was his investigation of Pacific flat-topped seamounts, for which he coined the name *guyot*. Hess considered that the tops of these seamounts had been planed flat by wave action at sea-level, and that they had since subsided by thousands of feet over a very long period; in fact the guyots were thought to be as old as the Pre-Cambrian. Consequently Hess was as surprised as anyone when Cretaceous fossils were dredged from the guyot tops in the post-war period, and more generally at the failure of ocean-wide dredging to yield rocks any older than this. Furthermore, the thickness of the sedimentary layer had proved less by something like an order of magnitude than predicted on the assumption that the ocean floor was ancient. Could it in fact be quite young, say no older than Mesozoic? The continental drift hypothesis of course predicted just this for the Atlantic and Indian oceans, but the Pacific was assumed to be much older, indeed primordial.

Finally, there was the discovery of the extensive mid-oceanic ridge system, with its seismicity, high heat flow, local volcanicity, and axial rift, implying tension.

Hess's great contribution was to integrate these disparate facts into the hypothesis that the mid-oceanic ridges are underlain by the hot, rising limbs of convection cells in the mantle, that the sea floor is being carried like a conveyor belt away from the ridge axes and thence under the marginal trenches by the cool descending limbs.[1] No wonder the ocean floor was young—it was constantly being renewed!

A special feature of Hess's hypothesis is that the oceanic crustal layer 3 is considered to be serpentinized peridotite, being in effect the hydrated top of the mantle (Fig. 21). Serpentine contains about 25 per

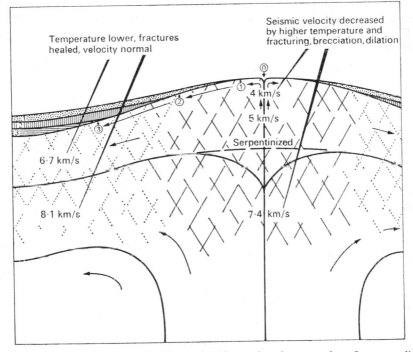

FIG. 21. Hess's interpretation of mid-oceanic ridge geology in terms of sea-floor spreading. After Hess 1962, Fig. 7.

[1] The term *sea-floor spreading* was in fact proposed not by Hess but by R. S. Dietz, whose name is often coupled with that of the Princeton geologist as one of the original propounders of the hypothesis. Dietz published a short paper in *Nature* in 1961, putting forward views very similar to those of Hess, whose lengthier work came out in print the following year. Hess's work had, however, been widely circulated in preprint form since 1960, and so there is no doubt about who has priority.

cent by volume of water, which could be derived by degassing of the rising column of a mantle convection cell. The remarkably constant thickness of layer 3 is controlled by the highest level reached by the 500° isotherm. (Above this temperature dehydration occurs.) As layer 3 is depressed into the downward-moving limb it will deserpentinize at 500°C back to normal peridotite and release water upward into the ocean.

It should be appreciated that the interpretation of layer 3 as serpentinized peridotite is not a vital part of the spreading sea-floor hypothesis. Layer 3 could just as well be (and in fact most probably is) basalt or gabbro derived from the mantle peridotite by partial melting.

Although we have been confining our attention to the sea floor, it is clear that here is a mechanism to cause the lateral displacement of continents. Rather than ploughing like so many ice-breakers through a weakly resisting sima, they are now seen merely to ride passively on the tops of conveyor belts. Precisely such a point was made over thirty years previously by Holmes, and comparisons of the two hypotheses are inevitable.

From our superior vantage-point in time it is not difficult to point out faults in Holmes's hypothesis. The crust–substratum model now seems primitive; eclogite cannot be produced by directed stresses but only under considerable lithostatic pressure; the East African Rift Valley has evidently been formed under tension, not compression, and so on. Some of Holmes's statements, furthermore, are vague and it is not always easy to pin him down to specific testable statements. Nevertheless, when all is said and done we must surely recognize his 1929 paper, based as it was on only a small fraction of what we knew by 1960, as a brilliant anticipation of the spreading sea-floor hypothesis. However, the contributions of Hess and Dietz were unquestionably more appropriate for their time and were readily testable with the new techniques available, as was soon to be shown.

Within a few years the spreading sea-floor hypothesis was utilized and elaborated in two different ways that were eventually to be integrated in a striking new theory.

J. Tuzo Wilson, of the University of Toronto, was impressed by the fact that movements of the Earth's crust were largely concentrated in three types of structural feature marked by earthquakes and volcanic activity, namely mountain ranges, including island arcs, mid-oceanic ridges, and major faults with large horizontal displacement. Especially puzzling was the fact that these features frequently seemed to end abruptly along their length.

In an article published in *Nature* in 1965 Wilson proposed that the mobile belts are in fact connected in a continuous network which divides the Earth's surface into several large, rigid plates. Any feature at its apparent termination could be transformed into either of the other two types. Such a junction was termed a *transform*. Faults on which the displacement suddenly stops or changes direction were termed *transform faults*. These are horizontal shear faults which terminate abruptly at

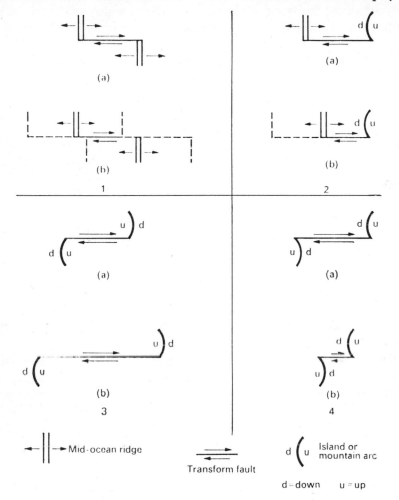

FIG. 22. Wilson's transform fault concept portrayed diagrammatically. *a*, earlier phase; *b*, later phase. The broken lines indicate former ridge positions. Based on Wilson 1965, Figs. 3 and 4.

both ends but may show great displacement of crustal blocks; they are not to be confused with transcurrent faults.

Fig. 22 illustrates some of the different types of dextral transform faults. As mid-ocean ridges expand to produce new crust according to the spreading sea-floor hypothesis, residual inactive traces of the former positions of the faults are left behind in their topography. On the other hand, as ocean crust moves down beneath the convex side of island

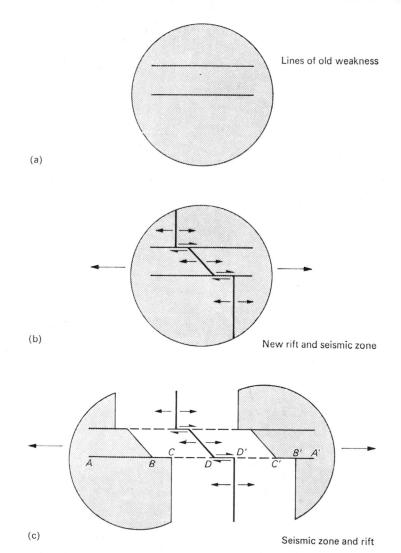

Fɪɢ. 23. Schematic portrayal of the opening of the Atlantic in terms of the transform fault concept. After Wilson 1965, Fig. 6.

arcs, the distance between two island arcs linked by a transform fault and convex towards each other will in the course of time diminish as crust is consumed. It is important to note that the sense of relative motion of crustal blocks is the opposite to that for a transcurrent fault.

Examine Wilson's simplified model of the opening of the Atlantic (Fig. 23). Parts AB and $A'B'$ of the east–west fault are older than the rifting, whereas DD' is young and in fact the only part now active, as shown by the location of seismic activity. The offset of the Mid-Atlantic Ridge is independent of the distance through which the continents have moved and merely reflects the shape of the initial break between the continental blocks.

Wilson cites as examples a number of famous faults which had hitherto been considered to exhibit normal transcurrent movement. Thus the San Andreas can be regarded as a transform fault linking the axis of the East Pacific Rise in the Gulf of California to the Juan de Fuca ridge south-west of Vancouver Island (Fig. 24). This neatly solves the problem of the northward continuation of the East Pacific Rise, which some had thought went directly under the interior plateaux of the western United States.

The new concept was quickly welcomed by a number of oceano-graphers who had been wrestling with problems associated with the offset of oceanic ridges and the disappearance of oceanic fracture zones at the margins of continents. Structural patterns on the ocean floor had begun to make a little sense.

The second line of approach concerned the interpretation of the puzzling patterns of magnetic anomalies described in the previous chapter, which since their initial discovery in the Pacific had been located in the other oceans as well and were evidently ubiquitous. After detailed mapping of the anomaly patterns on the Carlsberg Ridge and its flanks in the north-west Indian Ocean, a Cambridge research student called Fred Vine and his supervisor, Drummond Matthews, published a paper in *Nature* in 1963 which has come to be regarded as a landmark in the post-war progress of Earth science.

A short while previously palaeomagnetists and geochronologists working in collaboration had published data strongly supporting reversals of the geomagnetic field within the past few million years (see Chapter 4). When, therefore, the computed magnetic profiles assuming normal magnetization of the ocean floor were found to bear little resemblance to the observed profiles, the idea arose to try an alternative model with 50 per cent of the crust reversely magnetized in alternating bands. The reasoning behind this is as follows. If both sea-floor spread-ing and magnetic reversals occur, basalt magma would be expected to well up at the axis of the oceanic ridge, become magnetized in the

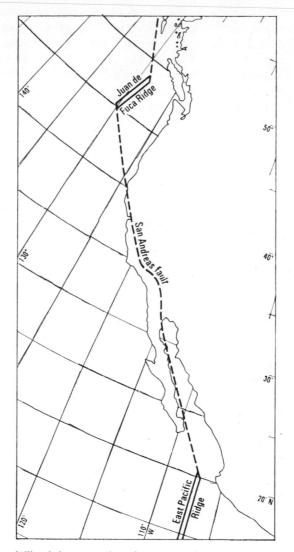

FIG. 24. Wilson's interpretation of the San Andreas as a transform fault.

direction of the geomagnetic field as it cooled below the Curie temperature to form a massive dyke, and spread laterally away from the axis.
Repetition of this process with the dyke being split axially would lead
to the formation of a series of blocks of alternately normal and reversed
magnetized material aligned parallel to the ridge axis, becoming
progressively older with increasing distance from it.

This time the computed profiles agreed very well with observation, as shown by Fig. 25. The 'Vine–Matthews hypothesis' could be said to be in business. It forms a nice example of what Koestler has called bisociation, an insight which arises from relating groups of facts or

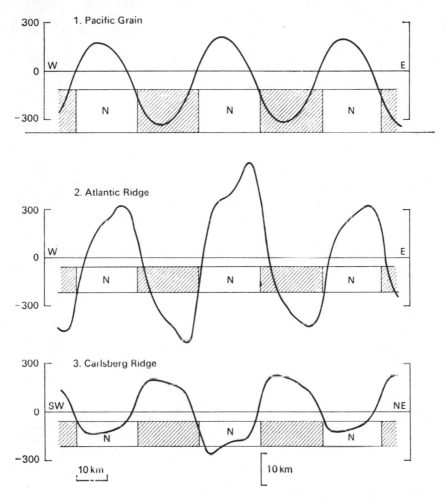

FIG. 25. Interpretation of magnetic anomaly patterns on the ocean floor in terms of magnetic reversals. N signifies normally magnetized; diagonally shaded signifies reversely magnetized areas. Oscillating lines are computed profiles. After Vine and Matthews 1963, Fig. 4.

ideas from apparently widely different fields, in this case Hess's concept and geomagnetic reversals. The implications were tremendously exciting. If Vine's idea were correct then we had a potential magnetic tape-recorder for determining the speed of the oceanic conveyor belt, given of course an accurate time-scale for reversals. Mapping of

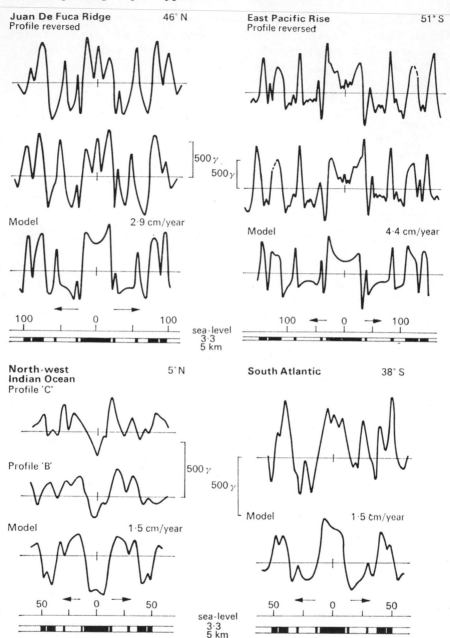

FIG. 26. Diagram to illustrate the close agreement between observed magnetic profiles and theoretical models based on the Vine–Matthews hypothesis. After Vine 1966, Figs. 6–9.

anomaly patterns on the ocean floor would allow much of the more recent history of the oceans to be inferred, such is the power of introducing a potentially quantitative technique.

But was the idea correct? Not many were convinced initially, and a few years were to elapse before conclusive supporting evidence was forthcoming. Two important requirements were to demonstrate that the anomaly patterns are exactly parallel with the mid-oceanic ridges and that they are symmetrical, the two sets of anomalies on each side of the ridges mirroring each other. This was first clearly demonstrated by W. C. Pitman and J. R. Heirtzler for the Pacific–Antarctic Ridge in 1966. Other ridges were soon shown to have similar characteristics (Fig. 26).

Meanwhile improved potassium–argon dating techniques for lavas had allowed the establishment of a more accurate geomagnetic reversal time-scale extending back over three million years. This agreed remarkably well with the sequence of reversals independently established by studying the magnetism of deep-sea sediment cores (Fig. 27). Vine realized that the rather distinctive sequence of normal and

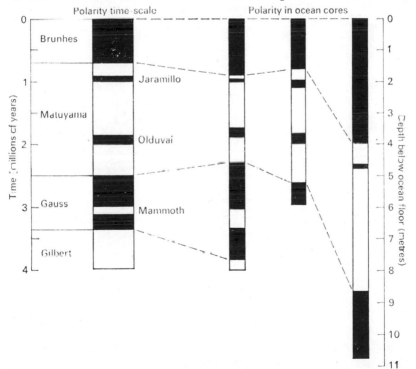

FIG. 27. Correlation of magnetic epochs based on polarity time-scale. Black, normal; white, reversed. Before becoming incorporated into sediments in ocean cores, magnetic particles become oriented in the direction of the geomagnetic field as they settle through the water. After Cox *et al.* 1967.

reversed polarity events was astonishingly similar to the sequence of magnetic anomalies on either side of the Juan de Fuca Ridge in the north-eastern Pacific.

The Vine–Matthews hypothesis could now be considered sufficiently probable to be taken as the basis of an intensive scientific programme (Fig. 28). In an important paper published in 1966 Vine reviewed the existing data on rates of spreading of the ocean floor. This rate (in effect

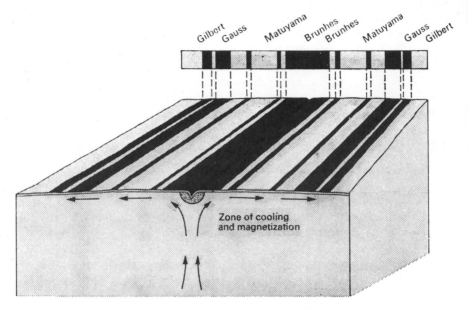

FIG. 28. Sea-floor spreading from the ridge axis in terms of the polarity time-scale (see Fig. 27). After Cox *et al.* 1967.

a half-rate, being concerned with only one side of a ridge) varied from about 1 cm/year on the ridge south-west of Iceland to as much as 4·5 cm/year in the south Pacific (Holmes had guessed at 5 cm/year). The spreading rates for the past few million years had been more or less constant for a given area. Extrapolation backwards in time at these rates would indicate an age for the ocean floor no older than Mesozoic, as Hess had indeed suggested. Active ridges could be distinguished from inactive ones by the presence of a pronounced central magnetic anomaly directly over the ridge axis. The reasoning behind this is that, over the course of time, scattered local volcanicity near the ridge axis, random with respect to normal or reversed magnetic fields, would tend to have the net effect of reducing the intensity of magnetization of the bands produced by upwelling and spreading of dykes. Only the central, freshly injected dyke would not be thus contaminated.

As if this were not enough for one year, the seismologist L. R. Sykes, working at the Lamont Observatory, was able to put Wilson's transform fault hypothesis to a critical test. Greatly improved seismic techniques made it possible to determine focal mechanisms on faults offsetting the oceanic ridges. Analysis of first ground motions at seismic stations throughout the world gave an indication of the sense of movement of one crustal block against its neighbour. Thus a clear-cut test became available to distinguish between transcurrent faults, which common sense had before Wilson's paper suggested as the obvious interpretation for the oceanic fracture zones, and transform faults, which imply sea-floor spreading. The new seismic data unequivocally supported the latter.

The year 1966 was undoubtedly one of breakthrough for the sea-floor spreading hypothesis. Before the year was out many Earth scientists began to take the idea of lateral mobility of continents seriously for the first time. Nowhere was the change in attitudes more dramatic than at the Lamont Observatory, which under the formidable directorship of Maurice Ewing had come to be without peer as an Earth sciences oceanographic institute. By all accounts the complete flip-over in beliefs of some of the leading personnel was as sudden and unequivocal as a magnetic reversal!

The next few years saw an intensive survey of magnetic anomalies over the world's oceans undertaken by researchers at the Lamont, notably J. R. Heirtzler, W. C. Pitman, and X. Le Pichon. Remarkable though it seemed, the same sequence of anomalies apparently occurred over the ridge flanks in different oceans and could hence be correlated with each other. This led to Heirtzler and his associates putting forward in 1968 a time-scale for the magnetic events back through the Tertiary, on the assumption of constant spreading rates. This bold extrapolation was quite naturally contested, but came through a critical test triumphantly before the decade was out.

This test depended on the latest exciting addition to the geological oceanographer's armoury of techniques. Financed by the U.S. National Science Foundation under a programme known as J.O.I.D.E.S. (Joint Oceanographic Institutes Deep Earth Sampling Programme) the drilling ship *Glomar Challenger* began a series of expeditions to the deep ocean with successive multinational teams of scientists on board. The ship was equipped with a unique, computer-controlled dynamic positioning system to maintain position in water too deep for anchoring. Sediment cores exceeding 1000 m could be obtained under a column of water over 6000 m deep—a remarkable feat of technology.

The third leg of the J.O.I.D.E.S. series of cruises involved a traverse across the South Atlantic at about 30°S. Drillings went down stratigraphically as far as the Upper Cretaceous and reached basaltic basement on a number of occasions. The age of the sediment directly overlying basement, as determined from microfossils, was found to increase systematically away from the axis of the Mid-Atlantic Ridge, implying a spreading rate of 2 cm/year, which agrees remarkably well with the Heirtzler time-scale (Fig. 29).

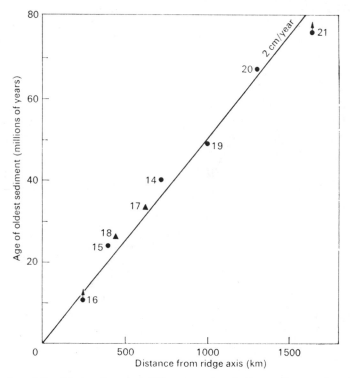

FIG. 29. Age of the oldest sediment at different numbered sites, as discovered on leg III of J.O.I.D.E.S., plotted against distance from ridge axis. Note the close agreement with a line based on a spreading rate of 2 cm/year. Adapted from Maxwell *et al.* 1970.

Subsequently the *Glomar Challenger* has undertaken many cruises. The volume of data accumulated from study of more than 200 drill cores is already enormous and should keep many scientists busy for a long time. Further core information has indeed tended to confirm the sea-floor spreading story in many details. The oldest sediments yet discovered overlying basaltic basement (or basaltic sills a short distance above basement) are Middle to Upper Jurassic, and occur in the north-west Pacific and the western North Atlantic. More than 50 per cent of the present ocean floor appears to have been created since the beginning of the Cainozoic Era.

The J.O.I.D.E.S. results suggest that the rate of spreading was more constant than was deduced earlier from observations of discontinuities in sediment thickness at the ocean ridge crest. This had suggested to M. and J. Ewing a sort of 'stop-and-go' spreading rate, with a far greater rate during the past ten million years than previously.

A further confirmatory test of the hypothesis came once more from seismology. The location of earthquake foci has long suggested a series of major faults under the Pacific marginal trench-island arc system dipping towards the continents on average at 45° and extending to depths as great at 700 km. Several independent studies, and notably one by J. Oliver and B. L. Isacks of the Lamont Observatory, established that the sense of motion on the fault-planes was such as to suggest movement of the oceanic crust downward beneath the trenches and arcs as Hess had proposed. The seismic data indeed showed that part of the upper mantle below the fault-zone had properties akin to the crust or topmost mantle further towards the middle of the ocean.

Before the decade had come to an end it seemed all over bar the shouting to any but the most diehard conservatives. The next major advance was to be the formulation of a new theory of global tectonics.

6. Plate tectonics

It is a popular delusion that the scientific enquirer is under an obligation not to go beyond generalisation of observed facts . . . but anyone who is practically acquainted with scientific work is aware that those who refuse to go beyond the facts, rarely get as far.

T. H. HUXLEY

THE GERMINAL idea behind the theory of plate tectonics is clearly present in Wilson's transform fault paper, as indeed is the first use in this context of the term 'plates'. Its full theoretical formulation and development was, however, due to three young men who had entered Earth science from physics; Jason Morgan and Dan McKenzie, of Princeton and Cambridge Universities respectively, and Xavier Le Pichon, who spent several years at the Lamont Observatory before returning to France.

Early in 1967 Morgan[1] had the idea of extending the transform fault concept to a spherical surface. He divided the Earth's surface into twenty blocks, some small, some large, divided by boundaries of three types: (1) ocean rises, where new crust was created, (2) trenches, where crust is destroyed, and (3) transform faults, where crust is neither created nor destroyed. In order to give the interpretative model mathematical rigour, the blocks were assumed to be perfectly rigid. The crust, especially the oceanic crust, is too thin to exhibit the required strength, and so the blocks or plates were thought to extend about 100 km down to the low-velocity layer of the mantle. The relatively rigid zone of the upper 100 km, termed the *tectosphere* by Morgan, has become more widely known as the *lithosphere*.

Euler's theorem states that a block on a sphere can be moved anywhere else on that sphere by a single rotation about a given axis. The velocity of relative motion of any two blocks is proportional to the angular velocity about the axis of rotation and to the angular distance from the axis. All faults common to the two blocks of Fig. 30 lie on small circles concentric about the pole of relative motion, and the velocity difference between the blocks (i.e. the spreading rate) increases as a sine of the angular distance from the rotation pole to reach a maximum at the 'equator'.

[1] In the same year McKenzie developed similar ideas with regard to the Pacific, which were written up with R. L. Parker in a paper in *Nature*.

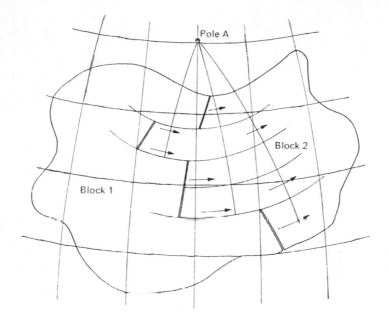

Fig. 30. Diagram to illustrate Morgan's application of Euler's theorem to lithospheric plates. After Morgan 1968, Fig. 4.

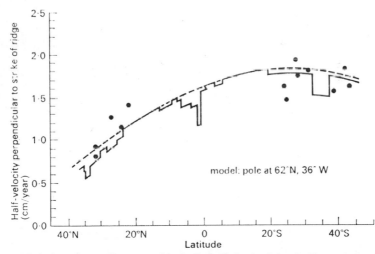

Fig. 31. Variation of spreading rate with the latitude in the Atlantic. Dots, observed rates; continuous line, predicted rates at right-angles to the ridge axis; dashed line, predicted rate parallel to the spreading direction. After Morgan 1968.

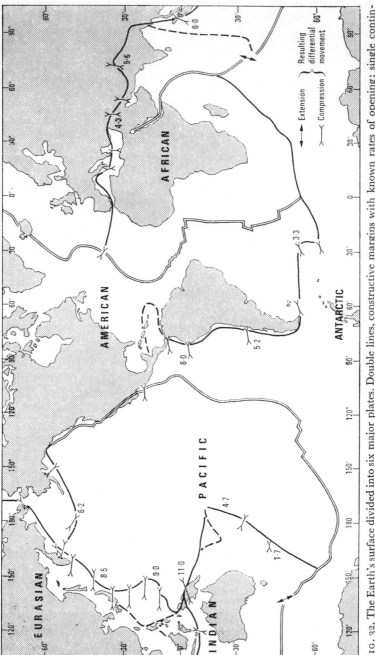

Fig. 32. The Earth's surface divided into six major plates. Double lines, constructive margins with known rates of opening; single continuous lines, destructive margins with computed rates of lithosphere consumption recorded in cm/year; dashed lines, boundaries of possible
minor plates. Adapted from Le Pichon 1968, Fig. 6.

Morgan found some factual support for such a change in spreading velocity down the Atlantic. The fracture zones in this ocean between 30°N and 10°S approximate to small circles with a pole near the southern tip of Greenland. The change with latitude of spreading rates estimated from the magnetic anomalies agreed with this (Fig. 31).

This original and rigorous approach to the Earth's surface quickly stimulated Le Pichon to attempt an ambitious survey of magnetic-anomaly and fracture-zone data over the whole globe. This time the model was even more simplified than Morgan's, with only six major plates distinguished, namely the American, Eurasian, African, Indian, Pacific, and Antarctic plates (Fig. 32). This disregards smaller plates

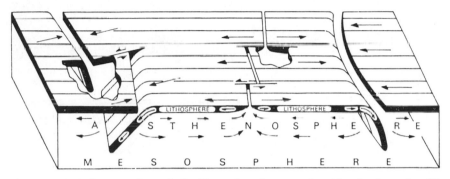

FIG. 33. Block diagram schematically illustrating plate tectonics. After Isacks *et al.* 1968, Fig. 1.

such as the Caribbean, which is thought to be a relict of Mesozoic Pacific crust that was emplaced between North and South America during separation from Europe and Africa, or the two plates in the eastern Mediterranean and Turkey whose boundaries mark the site of considerable earthquake activity.

Le Pichon used two independent sets of data to determine centres of rotation: the spreading rates deduced from magnetic anomalies and the azimuths of transform faults at the intersection with oceanic ridge axes. He succeeded remarkably well, considering the simplifying assumptions, in demonstrating a high measure of agreement between the two methods and showed how the opening of the South Pacific, North Pacific, Arctic, Atlantic, and Indian oceans can each be described by a single rotation. Apparently the plates do behave to a first approximation as rigid bodies.

Le Pichon rejected the notion that the oceanic spreading could be caused by a drastic increase of the Earth's radius in the last 200 million years, which indeed is implausible on a number of grounds. If there is no change in global radius, then crust created by spreading from the ridges must be consumed elsewhere in the 'sinks'. This allows estimates to be made of the rate of consumption below various trench–arc systems. We can at least in principle and to a considerable extent in practice fully determine the movement of lithospheric plates relative to each other since they are all interrelated. To give one instance, qualitatively the east–west vector of movement in the Atlantic opening can be treated as transformed by spreading along the Carlsberg Ridge into a north–south North Indian Ocean vector of movement towards the Himalayan 'sink'.

Finally, Le Pichon extrapolated spreading rates, that by now (1968) were known to reach 6 cm/year in the Pacific, backwards in time towards the Mesozoic in an attempt to unravel the history of oceanic opening.

In the same year that Morgan and Le Pichon published their papers the Lamont seismologists Isacks, Oliver, and Sykes pointed out with enthusiasm how much more successful the new global tectonics was than older explanatory models in accounting for earthquake phenomena (Fig. 33). Shallow earthquakes, with foci no deeper than a few tens of kilometres, characterize the tensional zones of the ocean ridges and also the movements along transform faults. Deep earthquakes with foci down to 700 km depth can occur only where compression forces the lithosphere to plunge down into the asthenosphere. What is especially striking is the relationship apparently established between the rates of underthrusting deduced from Le Pichon's analysis and the maximum depth of earthquake foci below the trenches. The directions of under-thrusting, furthermore, can be shown to correspond to the slip vectors derived from focal mechanism solutions.

W. Stauder and others attempted to account for the apparent paradox that a system under compression should nevertheless exhibit tensional phenomena in the ocean-floor sediments and shallow crustal zone. This was attributed to the sharp downward bending of the lithosphere causing the convex side of the bend to suffer tension. The lack of crumpling of sediments in some circum-Pacific trenches, as inferred from seismic reflection profiles, is still not properly understood. However, several possible explanations exist.

We may now summarize the basic tenets of plate tectonics and introduce a few more useful terms.

The Earth's crust and uppermost mantle down to a depth of approximately 100 km (usually somewhat more than this beneath the continents and less beneath the oceans) has significant strength in the sense of possessing enduring resistance to earthquake shear waves and is termed the *lithosphere*. Earthquakes, which result from the fracture of rocks under accumulated stress, are confined to this zone. The lithosphere overlies a zone of much weaker and hotter rocks, as marked by a significant reduction in shear-wave velocity, known as the *asthenosphere*, which is probably coincident, at least in part, with the low-velocity layer. The low resistance to shear stress is probably due to the fact that the rocks are close to their melting temperature. With increasing depth the rise of lithostatic pressure is more significant and the mantle viscosity, and hence shear velocity, increases. Thus the lithosphere is a kind of hard outer rind underlain by a more 'plastic' substratum.

Tectonic activity within the lithosphere has long been known to be largely confined to a series of narrow zones marked by earthquake and volcanic activity, most notably the 'Fiery Ring' around the margins of the Pacific. These tectonically active zones are now referable to the boundaries of six major, and about twice as many minor, more-or-less rigid, inert plates. The plate margins are of three types.

(1) *Constructive*, where new crust is created by the upwelling of material from the mantle, signified by the oceanic ridges.

(2) *Destructive*, where crust is consumed by being forced down beneath island arc–ocean trench systems or mountain belts, with the cold lithosphere descending down into the asthenosphere as far as 700 km. These commonly curvilinear features are sometimes known as *subduction zones* or *Benioff zones*, after the seismologist Hugo Benioff who emphasized the importance of the seismic zones indicative of faults inclined towards the continents.

(3) *Conservative*, where crust is neither created nor destroyed, the plates sliding laterally past each other. These are signified by the transform faults, which often offset the ocean ridge axes. The inactive continuations of transform faults beyond the offset ridge axes define circles of rotation for the previous motions and hence provide a key for the direction, though not the rate, of plate movements in the past.

Where two plate boundaries, or three plates, meet is known as a *triple junction*, which is indeed the only way that the boundary between two rigid plates can end. The geometry of triple junctions and the way in which they evolve, whether or not they maintain their location, have been analysed by Morgan and McKenzie. The boundary between the lithosphere and asthenosphere is a zone of decoupling. All plate boundaries may move with reference to a fixed geographic axis but the amount of crust created by spreading must balance the amount destroyed by thrusting down Benioff zones.

The amount of spreading away from ridge axes can be estimated indirectly from magnetic anomaly data, as described in the previous chapter. It may also be possible to estimate it directly, for instance across the central rift zone of Iceland, where scientists are currently using a laser-beam technique to measure the distance changes between widely spaced pillars over a number of years. The amount of crustal loss at destructive margins is inferred from computed spreading rates, utilizing the assumption that creation and destruction of crust over a whole globe of constant radius must balance each other. Judging from the record of magnetic 'stripes' on the ocean floor, plate tectonics has operated for close on 200 million years, and it is reasonable to extrapolate considerably further back in time than that.

It will be appreciated that the term *continental drift* is no longer strictly apposite because the plates may consist of both continent and ocean. Moreover, the crust–mantle division is less important than was previously thought, since the crust is coupled with the uppermost mantle. The important break is between the lithosphere and asthenosphere. Because of its buoyancy continental crust cannot be carried to depth. If therefore two continents collide at a destructive margin a thickening of the crust will take place and a mountain range will form. Thus the Himalayas can be seen as the result of India colliding with Asia over a Benioff zone. Alternatively a change in the pattern of plate boundaries will take place. This has probably happened in the case of the Red Sea and Gulf of Aden, which began to open up in the late Tertiary after Africa–Arabia had collided with Eurasia in the Middle East.

The question inevitably arises: what causes the plates to move? This is likely to be a speculative matter for a considerable time because of our ignorance of the mantle, which is bound to remain inaccessible to direct observation. Considerable progress can be made, however, towards understanding the driving mechanism by reducing the range of possibilities and constructing plausible models to fit the existing data.

In the early 1960s both Hess and Runcorn proposed models involving thermal convection within the mantle, with the migration of continents or ocean floor being coupled directly to the upper limbs of the underlying convection cells. This explanation fails to account for the phenomena of plate tectonics. Thus one fails to see how both the rising and descending limbs of convection cells can be abruptly offset by transform faults and it is hard to imagine the rates of motion of the rising limbs under the ridges increasing away from the axes of rotation in a manner corresponding to the plate motions. A further difficulty is that both Africa and Antarctica are surrounded on three sides by ridges but there are no destructive plate margins near these continents to consume the spreading sea floor. This poses no problem for plate tectonics because the plate boundaries can move with respect to each other, but a simple pattern of convection rolls of the type envisaged by Hess or Runcorn is clearly inadequate.

Recently argument has turned on whether the lithosphere is an active or a passive element in the dynamics. Elsasser has proposed that the cold descending slab of lithosphere beneath the island arcs is a major source of gravitational energy. Density contrast with the surrounding astheno-sphere will cause sinking and the sea floor will be pulled along in its wake, with mantle material welling up to fill the gaps created at the ridge axes. That the 'sinks' act as stress guides once they have been initiated seems very likely, but the hypothesis fails to account for their initiation, and it seems doubtful if motion could be maintained for long. Furthermore, one might expect the spreading rates on small plates surrounded by trenches to be greater than on large plates, but this is not so. On the other hand the median position of oceanic ridges and the symmetry of spreading is easier to account for by passive dyke-fill of the rifts than by active intrusion forcing the ridge flanks apart. Why otherwise should the youngest basalt always occur exactly in the centre?

The elevation of the ridges compared with the flanking plates is a further source of potential gravitational energy; thus the plates might partly slide under gravity. Stresses generated between small plates may also partly determine motions, but neither phenomenon can be envisaged as a major controlling factor. Density differences within the mantle are a different matter. Observations of gravity anomalies by orbiting satellites have provided information on these density differences, and they appear to be of the right order of magnitude to drive convection currents: about 1 to 10 cm/year.

Thermal convection of some sort seems to be the only sufficient source of energy. (This indeed applies to Elsasser's proposed mechanism, since the descending slab is seen as driven by a temperature-induced density contrast between it and the hotter mantle material through which it sinks.) Evidently, however, there can be no simple relation between mantle convection and surface features. Convection is most likely in the relatively hot and weak asthenosphere.

E. R. Oxburgh and D. L. Turcotte, working in conjunction, and D. P. McKenzie have mathematically explored possible convection models. What is envisaged is a kind of slow creep in crystalline rock. Though the concept is unfamiliar, such solids can indeed flow under certain conditions (consider, for instance, ice in a glacier). The rate of flow is enhanced by a high ratio of the actual temperature to the melting temperature and by low stress differences. If this temperature ratio is low, or the stress difference high, the rock will tend to fracture instead.

Emphasis is laid on what are termed *thermal boundary layers*. Heat is transferred by a kind of thermal conveyor belt including ascending hot and descending cold plumes, with the interior of the convection cell playing no part in the process. As the cold lithospheric slab descends down a Benioff zone a phase change to high-density minerals takes place, which enhances the sinking force. The slab eventually undergoes a kind of thermal erosion as it is warmed by conduction from the surrounding rock. Although the lithosphere is supposed to be cooled by the time it commences descent, island arcs are marked by high heat flow and abundant volcanicity. The necessary heat for rock fusion is thought to be provided by friction as the slab descends in the Benioff zone.

Morgan has drawn attention to so-called 'mantle hotspots', inferred from the evidence of volcanic island chains such as the Hawaiian islands. This string of islands, which become progressively older from east to west, is thought of as a series of volcanoes erupted from a hotspot fixed in position compared with the westward-spreading ocean floor. Iceland is an island lying astride the Mid-Atlantic Ridge with a considerably thickened oceanic crust compared with the rest of the ridge, composed principally of basaltic lavas. It may be envisaged as a hotspot symmetrically placed with respect to the ridge axis. None of the volcanic rocks on the island appears to be older than about 15 million years, but lavas in East Greenland and the Hebrides of Scotland have been dated at about 60 million years, the time of separation of Europe from Greenland. One North Atlantic hotspot that became active at the beginning of the Tertiary might therefore account not only for the observed pattern of volcanic activity but the plate movements as well. Morgan relates such hotspots to rising plumes of mantle material which start below the asthenosphere. Horizontal currents in the asthenosphere radiating from some twenty of these plumes, forming a 'thunderhead' pattern of convection, are sufficient to account for existing plate motions. The actual pattern of motions is, of course, determined by the positions of the plate boundaries.

Obviously there is no shortage of plausible suggestions for the driving force or forces. The difficulty lies in obtaining adequate supporting data and devising critical tests to decide between rival hypotheses. The equations governing the type of convectional flow likely to operate within the mantle, driven both by internal (radioactive) and shear-stress heating, are exceedingly complex, and numerical rather than analytical methods have to be used in such study. This is a field where we may reasonably expect considerable progress in the future.

7. The application of plate tectonics to continental geology

We see only what we know.

GOETHE

IT WOULD probably require a master of hyperbolic prose such as du Toit to do full justice to the impact which plate tectonics has had on the geological community at large. The published papers taking account of the new concept that began to appear from the close of the 1960s onwards give but a faint impression of the sense of excitement and expectancy felt collectively at meetings throughout the world but especially in North America and Great Britain. The scoffers and deriders of new-fangled disciplines and techniques were forced more and more on the defensive as scientific plums continued to fall off the tree. One must however reserve some sympathy for the more conservative geologists and palaeontologists.

By and large, classical geologists have been schooled in a tradition in which careful attention to observational detail and cautious inference of limited scope were accorded high status and the bold play of the imagination more or less subtly discouraged. There was a tendency to look askance at the flighty geophysicist who sometimes seemed to leap to the grandest conclusions on the flimsiest of evidence, only to abandon them just as dramatically a short while later. What was all the more galling, his arguments were so frequently wrapped up in pages of incomprehensible algebra. A caricature? Of course, just like the reverse view of a certain type of classical geologist as nothing more than a sophisticated stamp collector, but like all decent caricatures possessing more than a grain of truth.

The spectacular success of the geophysicists could, however, no longer be gainsaid by the great majority of geologists who had made any sort of attempt to keep in touch with the new developments. A series of bold hypotheses had emerged successfully from several critical tests. Much that had hitherto verged on the incomprehensible began to make real sense for the first time and a new coherent pattern of Earth evolution was dimly discernible. What had previously seemed swathed in complexity now looked as if it might be substantially explicable by reference to a few simple postulates. What geologist would have laid serious odds on the evidently remarkable resistance to distortion of most of the outer rind of the Earth, whose activities were to a first approximation reducible to the movement with respect to each other of a few thin plates?

The influence of plate tectonics was quickly felt in a number of fields and the new concept brought to bear on some traditional problems, such as the origin of mountain belts and the granitic intrusions, glaciations, and marine transgressions and regressions recorded in the stratigraphical record. The following examples have been chosen to illustrate some of the more active developments in the last few years.

1. Fitting together the components of Pangaea and timing their break-up

There had long been a need for a really accurate check on the supposed fit of the continents on the two sides of the Atlantic. Towards the middle of the 1960s two Cambridge scientists, J. E. Everett and A. G. Smith, undertook this task following a suggestion by Sir Edward Bullard that Euler's theorem might profitably be applied to the problem. We have already seen how this theorem was utilized by Morgan in his analysis of lithospheric plates. The technique involved the writing of a computer programme to determine the best least-squares fit of two irregular lines on a sphere. The lines to be fitted are digitized at numerous short geographic intervals. A 'homing' routine systematically searches for the best fit of the lines by rotating one of them about a rotation pole until it fits the other.

As for the lines selected, Everett found that the best fit for the South Atlantic was at the 500-fathom (approximately 1000-metre) contour. This is only a slightly better fit than at the 1000-fathom contour, which approximates more closely to the true edge of the continent, being about halfway down the continental slope.

The South Atlantic fit turns out to be remarkably good, with only trivial overlaps and gaps. What is more, the most important overlap, involving the Niger Delta, is irrelevant to continental drift because the delta is Tertiary in age and therefore younger than the postulated continental separation. (In a sense indeed this overlap could be held to support the argument.)

Recent geological work has confirmed the validity of du Toit's comparison of South Africa with South America, and added further evidence to underpin the geometric fit, such as a folded geosynclinal sequence of Pre-Cambrian strata in central Gabon which can be traced into the Bahia province of Brazil. Isotopic dating has, moreover, allowed the matching of ancient Pre-Cambrian 'shield' regions in West Africa and north-eastern Brazil, whose age exceeds 2000 million

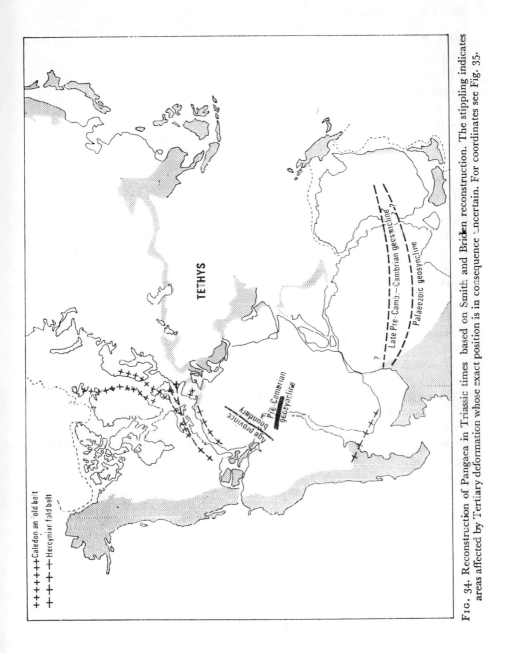

TETHYS

Late Pre-Camb.—Cambrian geosyncline ?

Palaeozoic geosyncline

Pre-Cambrian geosyncline

Age province boundary

FIG. 34. Reconstruction of Pangaea in Triassic times based on Smith and Briden reconstruction. The stippling indicates areas affected by Tertiary deformation whose exact position is in consequence uncertain. For coordinates see Fig. 35.

years, and a much younger zone of rocks to the south-east, with ages ranging from 450 to 650 million years (Fig. 34).

Matching the North Atlantic continents proved less straightforward. To get a reasonable fit of Africa with North America and Europe Smith had to assume that Spain had been rotated anticlockwise, opening up the Bay of Biscay, as Carey had suggested. It was also necessary to postulate that Iceland was no older than Tertiary and was underlain by oceanic crust, and that the submarine rise west of Scotland known as the Rockall Bank was a submerged part of the continental crust. All three assumptions have since been vindicated by geophysical and geochemical surveys, which is a triumphant confirmation of the method. The fit is also well in accord with the distribution of Caledonian and Hercynian mountain belts on the two sides of the present ocean.

On the other hand there has been some dispute about the closeness of fit of north-western Africa against the south-eastern United States. Such a fit, it has been argued, leaves no room for parts of central America which are known to be underlain by quite old continental rocks. One alternative is to widen the gap by fitting the opposing shelf edges along the boundaries of the so-called Quiet Magnetic Zones. These are marginal areas of the southern North Atlantic generally presumed to be oceanic crust, which occur landward of the zone of magnetic anomalies flanking the Mid-Atlantic Ridge. They were originally thought to be of Permian age or even older, and correspond to a time of no magnetic reversals. Subsequently, however, magnetically 'quiet' periods have been discovered in the Mesozoic. The most reasonable assessment today, based on recent J.O.I.D.E.S. drillings and the discovery of evidence of transcurrent fault displacements in central America, is that the computer fit is a close approximation to the truth, though one or two small misfits have to be accounted for.

The general excellence of the Atlantic fit of Bullard and his associates, based as it is on the application of Euler's theorem, provides of course an independent confirmation of the resistance to distortion of lithospheric plates.

A few years later the same technique was applied to the reconstruction of Gondwanaland by Smith and myself. The fit of South America and Africa was already unequivocally established, and that of Australia and Antarctica well underpinned geologically. Thus the late Pre-Cambrian and Cambrian Adelaide geosyncline of South Australia could be traced into the Trans-Antarctic Mountains which extend from the Ross to the Weddell Sea, while the Palaeozoic Tasman geosyncline of eastern Australia also found its continuation in Antarctica. Matching the other continental margins is more difficult.

The problem is to fit together the following pieces without leaving gaps:
(1) South America–Africa–Arabia, (2) Antarctica–Australia, (3) India. The only possible alternatives which did not do violence to the geology seemed either to fit India to Australia or to take India–Australia–Antarctica and fit it against South America–Africa–Arabia. In the latter case, which was adopted, a gap is left the same size as Madagascar.

Madagascar comes to lie against the coast of East Africa rather than South Africa and Mozambique, where it has sometimes been placed. There appears to be independent geological support for a more northerly location, as du Toit had first recognized. For instance, the strata of the late Palaeozoic Karroo System of Madagascar and Tanzania show close similarities and both countries lack the early Mesozoic volcanic rocks that are so widespread in southern Africa. There is also some oceanographic evidence to support the idea of southward drift of Madagascar relative to Africa.

Others prefer an alternative reconstruction whereby Madagascar is left in approximately its present latitude relative to Africa. There is some geological support for this also, for instance in the presence in Mozambique and Madagascar, but not in Kenya, of Cretaceous lavas which probably relate to tensional rifting. The difficulty about this reconstruction is that no adequate room is left for Antarctica, and the Antarctic Peninsula is forced towards the Pacific away from its obvious geological continuation in Patagonia, as established recently by I. W. D. Dalziel and D. H. Elliott.

Almost certainly there has been considerable movement between East and West Antarctica but at present we do not know how to allow for it. It therefore seems better to make a map in which Antarctica's shape is left unmodified, but needing change, than to try to guess what such a change might be. McElhinny's analysis show the relevant palaeomagnetic data to be consistent, furthermore, with the Smith–Hallam reconstruction.

The position of Madagascar is clearly the most enigmatic piece in the whole puzzle. Until more compelling evidence of a more southerly position comes forward than has hitherto been adduced, the Smith–Hallam reconstruction can be accepted provisionally as a reasonable approximation to the truth.

It will be noted that the reconstruction leaves a gap between India and Australia which corresponds with the Wharton Basin, lying in the Indian Ocean directly east of the Ninety East Ridge. This has been considered to be a possible segment of ancient ocean floor, presumably Palaeozoic, and indeed a thick series of Palaeozoic sedimentary strata are known in the western coastal regions of Australia. Recent

J.O.I.D.E.S. drilling has failed to confirm this. At least in its northern part, the basement of the Wharton Basin is no older than Cretaceous, and gets younger northwards, implying that a spreading ridge has been consumed down the Java Trench to the north. The Ninety East Ridge appears to be a transform fault.

The problem of dating the break-up of Pangaea and separation of its components can be approached on several grounds. Oceanographic data in the form of magnetic anomalies and deep drillings give probably the most accurate information, but the anomalies cannot be traced reliably beyond the late Cretaceous and many more drillings to basement will be required to give an adequate picture. Evidence from different fields of continental geology and palaeomagnetism, taken in conjunction, can already place the events in question within fairly narrow time-limits.

Thus the eruption of large quantities of basalts and other volcanic rocks near continental margins can reasonably be taken to relate to tensional fracturing associated with the severence of lithospheric plates. The advent of marine deposits in coastal regions with a long history of emergence is likely to signify either the proximity of newly created deep ocean, or at least subsidence of land below sea-level after crustal thinning because of tension, which may independently be inferred from normal faulting in coastal regions downwards towards the ocean. The marine strata may be preceded in time by salt deposits, signifying either limited access of sea-water to flat coastal plains in a warm, dry region, or brine concentration in a narrow strip of newly created deep sea with only restricted circulation with the open ocean. In terms of modern environments the former would be represented by the sabhka flats of the Persian Gulf or the Rann of Cutch in Pakistan, the latter by the Red Sea in the recent past. There is still dispute, however, about how supra-tidal and deep-water salt deposits may be distinguished.

The evidence of palaeontology should indicate at what time previously cosmopolitan continental organisms (and to some extent also shallow marine organisms) became isolated from each other, as marked by morphological divergence.

Close agreement in time between these different events obviously strengthens the interpretations of oceanic opening.

Comparison of the evidence from various fields suggests that there was sometimes a long time-interval (as much as tens of millions of years) between the first indications of tension, as shown by normal faulting, volcanicity, and crustal subsidence, and the subsequent dispersal of the continents by sea-floor spreading.

The first disruption seems to have taken place in the southern North Atlantic, and north-western Africa began to move away from North

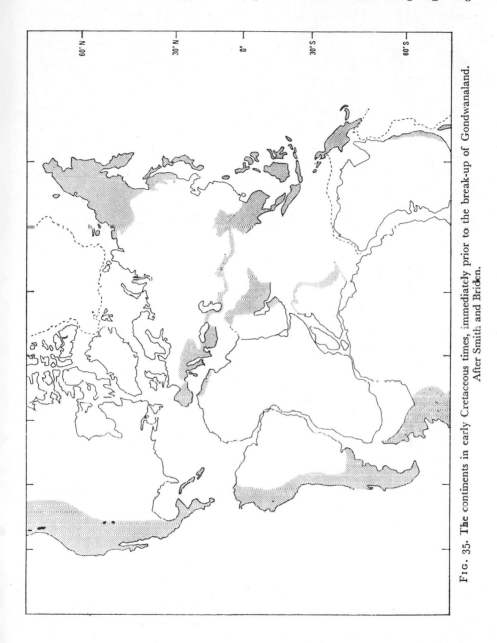

FIG. 35. The continents in early Cretaceous times, immediately prior to the break-up of Gondwanaland. After Smith and Briden.

America during the early Jurassic (Fig. 35). The zone of spreading moved progressively to the north and south. Africa and South America began to separate in the Middle Cretaceous, and Europe and North America started to move apart at about the same time. The opening of the North Atlantic appears to have been accomplished in several phases. Following the early separation of Africa from North America, Europe and Greenland broke away from Labrador in the late Cretaceous (about 80 million years ago) and the newly created Labrador Sea for a time formed the northern extension of the Atlantic. At 60 million years ago, approximately at the beginning of the Tertiary, the Rockall Plateau split off from Greenland, and from that time until the Middle Eocene both the Labrador Sea and North Atlantic between Europe and Greenland continued to widen. After the Middle Eocene (c. 47 million years ago), spreading ceased in the Labrador Sea but continued in the North Atlantic.

The Indian Ocean does not seem to have existed earlier than the Cretaceous; by the close of the period India was moving rapidly northward across the Tethys towards Asia, and Australia–Antarctica had parted from Africa, but sea-floor spreading between these two components did not begin until the early Tertiary, although tension between them, as marked by subsidence and marine transgression, had commenced at least as early as the Middle Cretaceous. A thick sequence of marine Mesozoic sediments off the coast of East Africa has been revealed by drilling for petroleum. These sediments are of shallow-water type and were probably laid down in a trough overlying a continental crust that had been thinned by tension before it was disrupted in the Cretaceous.

Dan McKenzie and John Sclater have undertaken a comprehensive analysis of magnetic anomaly data to deduce the evolution of the Indian Ocean since the Cretaceous. This turns out to have been highly complex; for example at least five different poles are needed to describe motion between India and Antarctica. India apparently moved northwards from Antarctica at the unusually high speed of 18 cm/year for some 20 million years in the early Tertiary. This rapid motion ceased in the Eocene and was followed by a period when little or no sea-floor spreading occurred in the western part of the ocean. Antarctica separated from Australia during this period.

It turns out from the analysis that the accurate reconstructions obtained by using the concepts of plate tectonics agree well with the latitudes established by palaeomagnetists. Further points of interest are that aseismic ridges in the Indian Ocean often mark plate boundaries and are formed during pauses in spreading, and that rifted continental margins are initially very steep and are gradually eroded.

2. The origin of mountain belts

Mountain belts are elongate linear or slightly arcuate features marked by distinctive zones of sedimentation and deformational patterns; the presence of extensive stratal folding and overthrusting suggests compression normal to their length. The sedimentary rocks characteristically exhibit signs of having been deposited at a rapid rate in deep water and typically consist of the sandstone–shale alternations known as 'flysch'. The clastic material of the sediments is frequently volcanic in origin and there may be intercalations of submarine 'pillow-lavas' or welded tuffs indicative of volcanic islands. This is a typical 'eugeosynclinal' association. Sometimes a thick sequence of shallow-water sedimentary rocks without volcanic material (character-istic of a 'miogeosyncline') occurs in a belt parallel and adjacent to the eugeosynclinal deposits, on that side nearer older rocks of so-called 'stable shield' areas.

It had long been recognized that the paired island arc–trench systems, with their intensive seismicity and volcanicity, possess many similarities and are quite probably mountain belts in the process of formation. There was more than a strong suggestion that, for instance, the mountainous islands of Japan belonged to an island arc–trench system that had been compressed and subjected to metamorphism and uplift in the late Mesozoic.

About a decade ago the Japanese geologist A. Miyashiro made the significant observation that the mountains of his country exhibit a pair of different metamorphic belts parallel to the length and adjacent to one another. On the side towards the Pacific are a group of meta-morphosed sedimentary rocks (schists), containing minerals indicative of formation at relatively low temperature but high pressure (e.g. glaucophane, aragonite, lawsonite). There is no evidence of any granitic basement. The other belt, to the west, does have granites and the associated metamorphosed sediments contain minerals such as sillimanite indicative of relatively high temperature and low pressure.

Subsequently such paired metamorphic belts, also formed during a late Mesozoic orogeny, were found elsewhere around the Pacific, for instance New Zealand and California, the glaucophane–schist (or 'blue-schist') belt always occurring nearer the ocean.

As interpreted by J. F. Dewey, J. Bird, W. R. Dickinson, and W. G. Ernst, the blue-schist belt is thought to have formed beneath ocean trenches, which appear to be the only places where the required temperature and pressure are likely to obtain. It appears quite probable that thick, intensely deformed sediment is scraped from the descending plate and plastered to the inner trench wall. Subsequent uplift exposes a so-called *mélange* terrain of complicated structure, in which shear surfaces supplant bedding as the dominant structural feature; the Franciscan rock-group of the Californian Coast Range forms a good example. The inner metamorphic belt, where a higher heat flow must have obtained, represents uplifted island arc; the Sierra Nevada–Klamath Mountains granitic complex and the associated schists and sediments are the corresponding Californian example (Fig. 39). The uplift of western North America is apparently a consequence of the overriding of the Pacific by the American plate during the Tertiary, which was in turn a consequence of the opening of the North Atlantic.

Another significant feature of most mountain belts are *ophiolite complexes*. This is the term given to masses of basic or ultrabasic igneous rocks such as basalt (often in the form of submarine pillow-lava), gabbro and peridotite, occurring as huge thrust-slices or slivers and especially characteristic of mélange terrains. The composition and structure of the rocks strongly suggests oceanic crust or topmost mantle, which has been sheared from downgoing plates and forced upwards under compression into the overlying rock. Commonly associated with ophiolites is the silica sedimentary rock known as radiolarian chert which, by analogy with modern oozes, is thought to be a deep-sea deposit.

Ophiolites provide an important clue to the presence of former subduction zones now enclosed by continent; in other words they are thought to mark the line of joining or suture of continents which have collided as a result of sea-floor spreading. Thus the ophiolite belt, showing thrusting of late Cretaceous age and extending from Oman through the Zagros Mountains of Iran and the Taurus ranges of Turkey, seems to mark either the closure of the African–Arabian continent on Eurasia or at least a subduction zone in the Tethys between Africa and Eurasia.

Dewey and Bird distinguish two types of mountain-building consequent upon plate movements (Figs. 36, 37). The first, island arc–Cordilleran type develops on leading plate edges above subduction zones and is marked by paired metamorphic belts, paired eugeosyncline–miogeosyncline relations, and divergent thrusting. The second, collision type forms after the impact of continent upon continent or continent upon island arc. There is no paired metamorphic zonation and the dominant metamorphism is of blue-schist type; the thrusting is dominantly towards and on to the consumed plate.

Both the Alpine–Himalayan and the circum-Pacific mountain groups relate to the plate movements which have caused Pangaea to disintegrate, but there is no reason to believe that older mountain belts formed in a different way. Dewey has, for example, followed up a proposal by Tuzo Wilson that the early Palaeozoic Caledonian

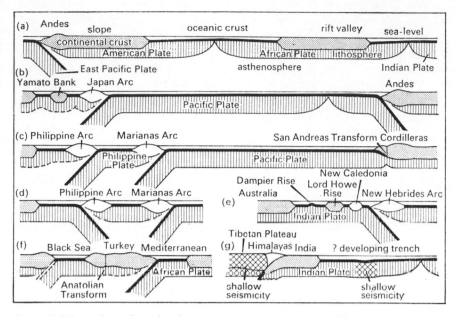

FIG. 36. Schematic sections showing plate, ocean, continent, and island–arc relationship. After Dewey and Bird 1970, Fig. 2.

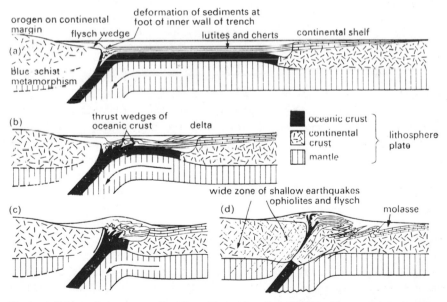

FIG. 37. Schematic sequence of sections illustrating the collision of two continents. After Dewey and Bird 1970, Fig. 13.

mountain belt of north-west Europe and eastern North America marks the line of closure of an ancient ocean. By synthesizing a wide variety of data of the type outlined in the preceding paragraphs he has argued persuasively for an old line of continental suture extending from southern Scotland (Fig. 38) to central Newfoundland. Likewise, W. Hamilton has presented a convincing case for the Ural Mountains being a late Palaeozoic line of closure between Europe and Siberia. The solitary supercontinent Pangaea evidently did not exist earlier than about 300 million years ago.

Orogenic belts evidently give significant clues to the creation and disappearance of oceans throughout long stretches of geological time. We should beware, however, of falling into the complacent assumption that we necessarily have the final key to understanding mountain belts. Many geologists still believe that Pre-Cambrian orogenic belts may be fundamentally different in type from younger ones formed in the last few hundred million years. There is no reason why this should not be so if, as widely believed, the crust was appreciably thinner in the early Pre-Cambrian, so that a different pattern of stresses might have operated.

Perhaps more embarrassing, no one has yet satisfactorily accounted for the late Palaeozoic Hercynian orogeny in western Europe. No paired metamorphic belts nor extensive ophiolite complexes are known, and the zone of stratal folding and thrusting, and metamorphism and igneous intrusion, extends over a very wide area not easily transformable into a narrow belt even by the most imaginative reconstruction.

Finally, not all mountain belts, as marked by zones of compression and uplift, necessarily form at plate margins.

3. Igneous activity

The eruption of basalt in the oceans is readily explicable in terms of plate tectonics. It has been known for some time from experimental work that an ultrabasic rock like peridotite or a hypothetical near-equivalent called pyrolite will yield on partial melting a magma of basaltic composition. Careful monitoring of volcanic eruptions in Hawaii has revealed that eruptions are normally preceded by minor earthquake shocks originating in the upper mantle, indicating the likely source of the lava. All that is required is a heat source from below, such as can be provided by a rising plume of hot mantle material, either under a ridge or a 'hotspot', and the local existence of tension in the crust. Eruption of basalt lavas on the continents likewise demands a source in the upper mantle and tension in the overlying crust. Gabbro and dolerite are simply coarse-grained equivalents of basalt *intruded* at depth in the crust rather than *extruded* at the surface as a lava and

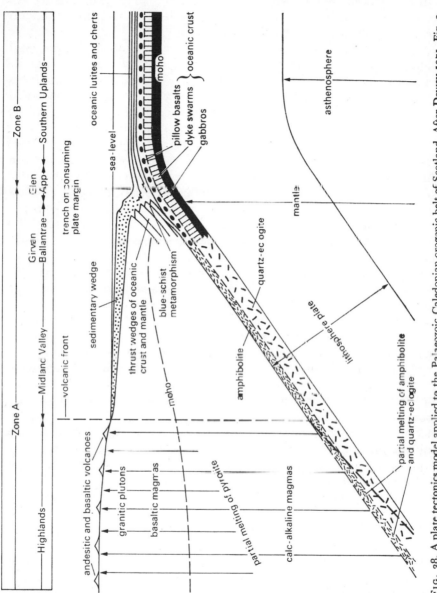

F1G. 38. A plate tectonics model applied to the Palaeozoic Caledonian orogenic belt of Scotland. After Dewey 1971, Fig. 2.

hence cooled more quickly, with the consequent formation of small crystals.

The origin of andesitic volcanic rocks and the huge granitic intrusive bodies known as batholiths, which are confined to the continents, has been a far more controversial matter. Both groups of rocks are lighter in colour and weight, richer in silicon, aluminium, sodium, and potassium but poorer in calcium, magnesium, and iron than basalts and gabbros. Experimental evidence clearly indicates that such rocks can form as late-stage, relatively low-temperature products of chemical differentiation of a parent magma of basaltic composition. However, the huge bulk of granitic batholiths, containing rocks so rich in silica that free quartz may comprise upwards of 10 per cent of the minerals present, has been widely held to rule out magmatic differentiation of this sort. Instead, because their chemical composition is closely similar to a sandstone–shale mixture or its metamorphic equivalent, granites and related rocks have been thought by many to originate from the melting of sediments deep in the crust, or even, in an extreme view, by the diffusion of ions through rocks without the formation of liquid magma.

The new concepts of plate tectonics have now been critically brought to bear on these problems by W. R. Dickinson and others. What may be termed *andesitic* lavas, but ranging in composition more widely than andesite, characteristically form elongate volcanic island arcs today, and are usually preserved as arcuate belts in the past. *Granitic* 'plutons' or intrusive masses (mainly granodiorite) also occur around the Pacific as elongate, curvilinear batholiths slightly farther towards the continental interior. Both the physical relationships of the two types of igneous bodies and the close similarity in chemistry, with the granitic rocks only a little richer in silica, suggest that they are related; that in fact both come from the same magma-type, one group being extrusive and the other intrusive. The batholiths are thought to form at some depth in island arcs and to be subsequently exposed by erosion biting through the uplifted mountains such as the Californian Sierra Nevada and Peruvian Andes. Independent evidence summarized in the previous section supports the notion that these are uplifted island arcs.

Various lines of argument point against a deep continental crustal origin for the rocks in question. Some andesites are known to be underlain by oceanic crust, and their trace-element distributions are consistent with derivation from basalt. A temperature of 1000°C or more, such as is required for rock fusion to produce the right sort of magma, is now thought unlikely to be reached within the continental crust. The ratio of the strontium-87 and 86 isotopes in both the andesitic and granitic rocks suggests derivation from oceanic material with minimal contamination by the continental crust. The linearity of the batholith belts and the episodic character of intrusion do not favour an origin by melting of continental crust.

In the plate tectonics model, which seems best to fit the data, andesitic magma is generated by differentiation or partial melting, or both, of descending slabs of lithosphere, the heat of fusion largely being produced by friction with the overlying plate. As the slab descends, basalt eventually converts under increasing lithostatic pressure to eclogite, which reinforces the downward pull because of a higher density than peridotite. It has been established empirically that the content of potash increases directly with the depth of the seismic zone in present-day island arcs. This probably relates to different crystal-melt equilibria at different temperatures and pressures. Varying potash contents can likewise be traced over the ancient island arcs represented today by granitic batholiths, and are held to give a clue as to the local depth of the Benioff zone (Fig. 39).

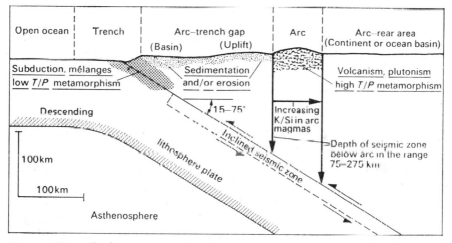

FIG. 39. Generalized transverse section of an arc–trench system showing relation of magmatism and metamorphism to the subduction zone. After Dickinson 1971, Fig. 3.

The whole range of clastic sediments, from quartz sandstones to greywackes with volcanic fragments, and shales, can readily be derived from these igneous rocks. With the somewhat surprising conclusion that granites as well as andesites were apparently derived more or less directly from the oceanic lithosphere a ready mechanism presents itself for the building of continents, the present tectonically stable shield areas having originally been island arcs and trenches long ago in the Palaeozoic or Pre-Cambrian. It has been estimated that, extrapolating from rates of eruption known in the fairly recent past, the whole of Japan could have been formed by the mechanism outlined in 3000 million years, which is considerably less than the age of the Earth, and the near-by Kurile Islands in a mere 75 million years.

4. Vertical tectonic movements and sea-level changes

While plate tectonics has been highly successful in accounting for lateral movements of the crust, it remains much less clear what causes the regional uplifts and subsidences known as epeirogenic movements that occur both on plates and plate margins. Even the classic problem of mountain-building, in so far as mountains have to be uplifted, has not been fully accounted for by the new concept. Under what circumstances, for instance, does an island arc–trench system become converted into a mountain belt? Though continent–continent collision (with underthrusting) is supposed to have caused uplift of, for instance, the Himalayas, why has there been a long time-interval between collision and the substantial uplift from late Tertiary times onwards? What indeed, more generally, are the time-relations between lateral and vertical movements in mountain belts? More obscure still, what controls epeirogenic movements in so-called stable shields or cratons, which are supposed in plate theory to be tectonically inert?

Substantial progress has nevertheless been made towards understanding the control of uplift of the mid-oceanic ridges and subsidence of 'stable' continental margins such as occur around the Atlantic and Indian Oceans. Deep boreholes have penetrated thousands of metres of shallow-water sediment on the continental shelves of these oceans, proving more-or-less steady subsidence since the late Mesozoic.

M. G. Langseth and others have studied the variations of ocean-floor heat flow and found that it decreases systematically with increasing age of the ocean floor away from the ridges. As the ridges are isostatically compensated this suggests that their relatively high relief is due to thermal expansion of mantle material. Supporting evidence is provided by the abnormally low seismic wave transmission velocities found beneath the ridges (Fig. 18).

J. G. Sclater and J. Francheteau have demonstrated for the North Pacific that the actual topography of the sea floor approximates quite closely to a theoretical model based on heat-flow data and estimated thermal expansion of mantle material (Fig. 40). Agreement might be even closer if more were known about the mantle; small variations in the high-pressure and high-temperature fields of certain mineral assemblages in mantle rocks would easily yield a distribution of density contrast that would give a better fit. Heat-flow data for other oceanic areas are more sparse but indicate that the model may have general validity.

Progressive subsidence of the ocean floor away from the ridge axes is

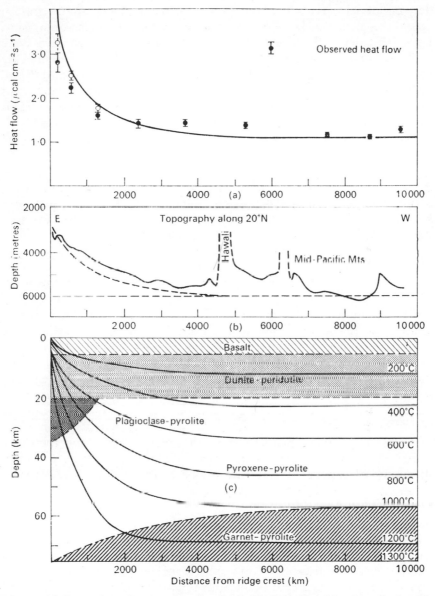

FIG. 40. (a) Comparison of average heat flow in the North Pacific with a theoretical profile for a lithosphere 75 km thick. (b) Comparison of observed topography (continuous line) with theoretical profile based on thermal expansion model. (c) After Sclater and **Francheteau** 1970, Fig. 17.

also indicated by data from the J.O.I.D.E.S. boreholes. Present-day deep ocean deposits above about 4500 m below sea-level usually contain abundant calcium carbonate in the form of skeletons of micro-organisms. Below this depth the effect of increased pressure is to dissolve these skeletons, and the bottom sediments are represented by inorganic clay or by oozes containing only the remains of siliceous micro-organisms. Some of the J.O.I.D.E.S. drillings have shown a change from calcareous to non-calcareous deep-sea ooze up the stratigraphical sequence. This can most reasonably be interpreted as due to subsidence below this critical *compensation depth* as the ocean-floor spread away from the ridge axis.

It also follows that some at least of the subsidence of the Indian and Atlantic ocean margins is likely to be the consequence of cooling of the ocean floor. An additional factor, suggested by M. H. P. Bott, may be a kind of creep of the trailing edge of the severed continental crust towards the new ocean, leading to thinning and hence subsidence.

One of many interesting suggestions made by Holmes in his 1929 paper was that the world-wide marine transgressions might relate to uplift of oceanic ridges during continental drift. Exciting confirmation of this possibility comes from the work of Tj. Van Andel and G. R. Heath on the Mid-Atlantic Ridge south of the equator. They infer a number of uplifts and subsidences of the ridge during the Tertiary of a magnitude sufficient to cause significant displacement of sea-water on to or off the continental margins. Evidence is cited to support an interesting correlation between uplift and acceleration of sea-floor spreading rate, and vice versa. This might mean, for instance, that the major Upper Cretaceous transgression recorded in the stratigraphical record of all continents was a consequence of the increased spreading rates inferred on independent grounds. Alternatively, the increased area of the uplifted ridge system which is implied by the complete disruption of Pangaea during that period could be the prime cause.

5. A major rift system

One of the geologically most fascinating regions in the world is where the Ethiopian, Red Sea, and Gulf of Aden rift systems meet. This was the subject of a Royal Society symposium in January 1970 in which a number of geologists and geophysicists reported their findings on land and at sea and interpreted them in the light of modern concepts.

Both the Red Sea and Gulf of Aden are underlain by young oceanic crust and are now generally believed to have opened up by sea-floor spreading in the late Tertiary, starting in the Lower Miocene. The Red Sea rifting was preceded in the early Tertiary by domal uplift of the adjacent parts of the Nubian and Arabian shields, so that a roughly oval area of ancient Pre-Cambrian rocks is surrounded by younger rocks in concentric zones. The separation of Arabia from Africa was accompanied by transcurrent movement along the Dead Sea fault-zone of the Near East, and by movement along transform faults in the Gulf of Aden (Fig. 41). The principal movements can be described with

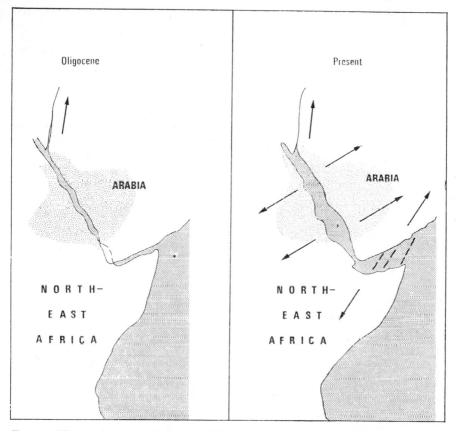

FIG. 41. The opening of the Red Sea and Gulf of Aden. Stippled zone is approximate area of the Arabo–Nubian Pre-Cambrian shield; dashed lines in Gulf of Aden are transform faults; arrows indicate principal movement vectors.

respect to a rotation pole situated near the north-east Libyan coast, but the detailed structural pattern is complex and a number of secondary vectors of movement have to be invoked, such as the slight clockwise rotation of the Horn of Africa.

Fig. 41 shows that if Arabia is closed up on Africa the general fit is excellent but an area of overlap exists in the triangular-shaped Afar depression. Geophysical data indicate that the crust of this region has a structure intermediate between normal continental and oceanic. It is thought to represent an area of extended, thinned, and fragmented continental crust intruded by dense, basic igneous rocks and covered by a series of late Tertiary basaltic lavas. There is still some dispute about how wide the true ocean floor is in the Red Sea. Some would argue that it is considerably narrower than is portrayed in Fig. 41.

As is to be expected in a region of crustal tension, volcanic rocks are extremely abundant. I. G. Gass has established an interesting correlation between the lavas and the style of tectonic deformation, which throws light on the factors controlling the origin of different types of magma. Three phases are distinguishable. The first phase was marked by the eruption of vast quantities of alkali basalt in Ethiopia, the Yemen, and southern Arabia, preceding and concurrent with regional tectonic uplift in the early Tertiary. These basalts are characterized by a high ratio of sodium to silicon and of ferric to ferrous iron and a low titanium content. The second phase correlates with continental crust attenuation and lateral movement in the Miocene; it is marked by so-called peralkaline volcanics including silica-rich differentiates.

The third phase is associated with the opening of the Gulf of Aden and Red Sea. Magnetic anomaly patterns and other geophysical data indicate that these seas are indeed juvenile oceans created by the welling-up of basalt along their axes and by spreading away from these as the continental crust was disrupted by tension.

The lavas of the third phase are of true oceanic type, that is, tholeiitic basalts relatively rich in silicon and calcium, but low in titanium and very low in potassium.

Another interesting discovery from drilling the sediments underlying the Red Sea is that of salt deposits of Miocene and younger age. These occur under fairly shallow water and were deposited on continental crust. They are thought to have been deposited in marginal basins restricted from free marine circulation as a result of the asymmetrical fault-collapse of continental crustal blocks associated with the tensional effects of opening of the Red Sea.

In so far as these deposits form an interpretative model for the similar late Triassic and early Jurassic salts margining the southern North Atlantic, and are thought to mark an early 'Red Sea' phase of Atlantic rifting, they support the shallow- rather than the deep-water interpretation of salt deposition. Certain deeps within the Red Sea do, it is true, contain hot, highly saline and metal-enriched brines, but they are not underlain by salt deposits. This renders acute the problem of undoubted salt diapirs under the deeps of the Gulf of Mexico. If the salts were laid down in deep water, as had been suggested, where is the actualistic analogy? On the other hand, if they are in reality shallow-water deposits that have subsided to great depths subsequently, where has the continental crust gone, because the crustal structure under the deeps of the Gulf of Mexico is suboceanic?

6. Former ice ages

Wegener had been troubled by the widespread agreement among his contemporaries that a certain Carboniferous boulder-bed near Boston, New England was glacial in origin. Indeed it was termed the Squantum Tillite. As it occurred within the heart of his postulated tropical zone for that time it proved something of an embarrassment. About a decade ago, however, the supposed tillite was carefully reinvestigated in the light of modern sedimentological knowledge by R. H. Dott and found not to have any of the key characteristics of a glacial deposit. Wegener's claim that it could not be tillite because the associated coal beds indicated otherwise appears therefore to have been posthumously vindicated.

Two American geologists, J. C. Crowell and L. A. Frakes, have over a period of years made a detailed investigation of the sedimentology and distribution of the late Palaeozoic glacial deposits of South America, South Africa, Australia, and Antarctica. They have inferred directions of ice movement from such features as striated pavements beneath tillites, the orientation of stones within the tillites, and various morphological features of the subglacial topography. Lateral passage of ancient glacial boulder-clay into outwash sands and thence subaqueous deposits have been traced to give indications of regional slope.

During the Lower Carboniferous, when according to the palaeomagnetists the pole was in the Transvaal, glaciers radiated from several centres in southern Africa and perhaps westward into the Paraná Basin of South America; there was also some local glaciation in South America and Antarctica.

The glaciation reached its maximum extent in the Upper Carboniferous (Fig. 42), when the pole was in west Antarctica. A lobe of the principal ice-mass, which covered most of southern Africa, extended into the Paraná Basin and ice probably covered a large part of Antarctica. An ice-cap occupied Victoria Land in this continent, and glaciers extended from here into Tasmania and South Australia; a separate ice-cap was formed in the middle of India. From the evidence of marine deposits Crowell and Frakes infer an inland sea extending as an embayment of the ocean in the area between Antarctica, Africa, and South America. If unfrozen this could have provided a source of moisture for the inland ice.

In Lower Permian times the pole had moved closer to Australia and glacial centres had nearly disappeared from all but here, Antarctica,

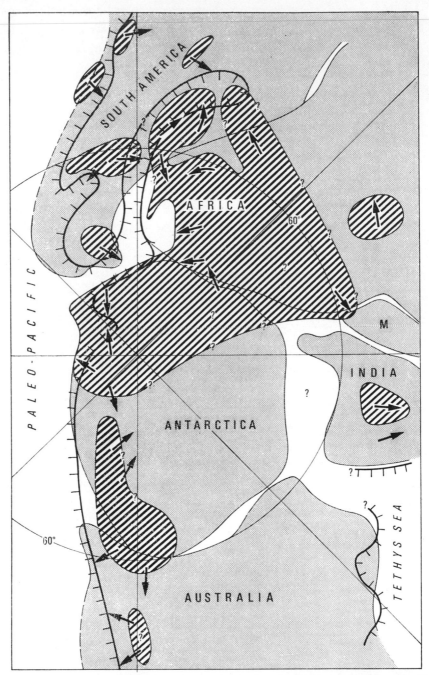

FIG. 42. The southern hemisphere Ice Age in the late Carboniferous. Stippling signifies continent, hatching ice-caps, and arrows direction of ice movement. After Crowell and Frakes 1970, Fig. 3.

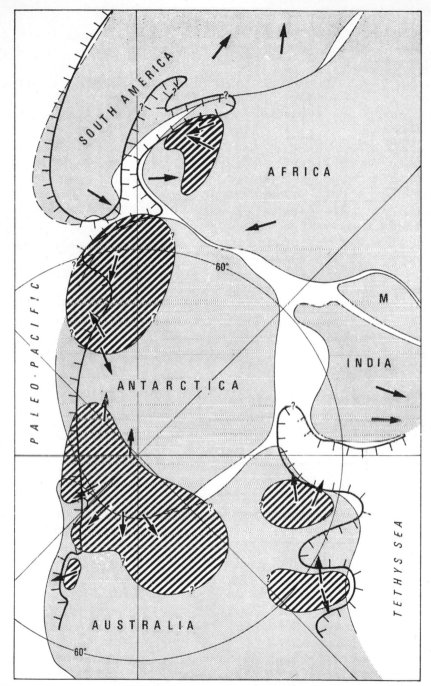

FIG. 43. The southern hemisphere Ice Age in the early Permian. Symbols and ornaments as for Fig. 42. After Crowell and Frakes 1970, Fig. 4.

and a small area in South West Africa (Fig. 43). The glaciation of Australia, experienced both in the east and west, was more extensive than previously.

A few years ago French geologists discovered evidence in the Sahara of a major Ordovician glaciation, which strikingly confirms the palaeomagnetists' estimate of a pole position in North Africa at about this time.

More generally, it is desirable to examine M. Ewing and W. L. Donn's celebrated hypothesis of the origin of ice ages in the light of plate movements. This hypothesis propounds that ice ages result from movement of the poles into thermally isolated positions. The equable pattern of climate, for instance, in the Mesozoic and much of Cainozoic time can be attributed to pole positions in the open ocean; there is every indication that no permanent ice cover could form in such circumstances, even with the same pattern of solar radiation as in the Pleistocene. The late Cainozoic Ice Age resulted, according to the Ewing–Donn hypothesis, when the south pole moved into the Antarctic continent and the north pole into the almost landlocked Arctic Ocean.

If the hypothesis is correct, a high concentration of land would be expected in high latitudes during ice ages. The geological and palaeomagnetic reconstructions of Gondwanaland in the late Palaeozoic indicate just this.

7. Ancient faunal distributions

Although the distribution of fossil organisms had been considered by Wegener to provide vital support for his hypothesis, the initial response of palaeontologists to the new developments in Earth science was not notably sprightly, but momentum is now rapidly building up and the research field of palaeobiogeography gaining many new adherents.

Continental vertebrates have always been considered to be especially reliable indicators of land connections because of their general inability to cross major sea barriers. A few groups such as rodents, bats, and frogs can be inferred to have crossed fairly narrow marine straits spasmodically, and island faunas, characterized by low variety and poor ecological balance, may have originated from occasional chance crossings on driftwood. (Birds can be ignored since they have a negligible fossil record.) The identity or near-identity of whole faunas between different continents must, however, be accepted as evidence of free land communication.

Consider the biogeographical patterns exhibited by mammals, which radiated into a wide variety of ecological niches and expanded greatly in population and size after the extinction of the dinosaurs and other reptiles at the close of the Mesozoic. Everyone is familiar with the distinctive marsupial–monotreme fauna of Australia, which developed in isolation from the more advanced placentals that elsewhere quickly established their dominance. What is less well known is that during most of the Tertiary, until in fact the uplift above sea-level of the central American isthmus, South America had just as distinctive a fauna as Australia, with such groups as notoungulates, liptoterns, and glyptodonts. These became extinct in the Pleistocene, presumably as a result of invasion of their land by more successful North American mammals.

The Tertiary faunas of Africa also had distinctive elements indicative of at least partial isolation, such as the ancestors of our living elephants, conies, and sea cows.

All these facts are consistent with the currently accepted model of continental drift, whereby the three continents in question became isolated from each other and from the northern continents by sea in the late Mesozoic or beginning of the Tertiary, allowing genetic isolation and hence morphological divergence.

In the northern hemisphere the mammals of North America and Europe were closely similar in type until the early Eocene, after which they diverged sharply. This is consistent with sea-floor spreading data suggesting that a land connection persisted in the north until about this time. The earliest mass migrations of mammals between Africa and Eurasia seem to have taken place early in the Miocene, a time when geological evidence suggests that Africa–Arabia had collided in one or more places with the northern landmass. Incidentally, this marks the time when free communication of marine invertebrate faunas through the Tethys from the Caribbean to southern Asia was interrupted for the first time, thus providing a neat means of cross-checking the interpretation of land migrations.

When we turn to the Mesozoic, when reptiles were the dominant land vertebrates, a totally different picture emerges. Whereas the Cainozoic mammals fall clearly into several distinct geographic provinces, no such provinces can be distinguished for the Mesozoic. Where the fossil record is relatively good, as in the Triassic, a cosmopolitan fauna is recognizable, implying free land communication between all the continents. The same pattern seems broadly true also for Jurassic and Cretaceous dinosaurs. Three genera of especially huge animals are recorded from Upper Cretaceous strata in several southern continents and India. If we can trust the identifications, this argues strongly for postponement of Gondwanaland disintegration until well into the Cretaceous, as is suggested independently by geological and geophysical evidence.

Quite evidently the making and breaking of land connections at different times by continental drift must have an important bearing on interpretations of evolution and extinction patterns, although the relationship has as yet hardly begun to be explored thoroughly. Another interesting biological approach to the fossil record in the light of plate movements is to study the *diversity* or taxonomic variety of whole faunal assemblages. This is best done with marine invertebrates, which have left a prolific fossil record.

A palaeontologist, J. W. Valentine, and a structural geologist, E. M. Moores, of the University of California have collaborated to produce a stimulating article suggesting how invertebrate diversity patterns may relate to plate tectonics since the Pre-Cambrian. As portrayed in Fig. 44, the number of families is seen to increase sharply from the Cambrian

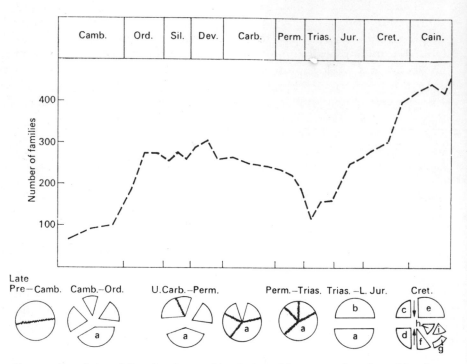

FIG. 44. Correlation of diversity changes through time with patterns of continental break-up and suturing. *a*, Gondwanaland; *b*, Laurasia; *c*, North America; *d*, South America; *e*, Eurasia; *f*, Africa; *g*, Antarctica; *h*, India; *j*, Australia. Adapted from Valentine and Moores 1970, Fig. 1.

to the Ordovician and to drop just as sharply in the late Permian. There is subsequently a strong increase in the Mesozoic, which becomes especially marked in the late Cretaceous.

Since most invertebrate species live as adults on the continental shelf, and since the larvae of many of them cannot survive the time taken to cross a deep ocean, drifting apart of continents should lead to genetic isolation and hence new species formation on a large scale, though less marked than with land animals. Hence continental fragmentation should have the net result of increasing world diversity. Furthermore, a series of small continents would result in more equable climates and greater environmental stability for the shelf organisms, which also favours increased diversity according to current ecological theory. On the contrary, a solitary supercontinent would result in extreme seasonal contrasts in climate of the shelf seas, which should reduce diversity. Suturing of formerly separated continents would cause overlap of ecological niches by previously isolated groups of organisms and the ensuing competition should result in reduction of diversity through mass extinction. Fig. 44 indicates an interpretation along these lines. It should be noted though that the Palaeozoic structural interpretation is rather speculative.

These examples from several widely different research fields should indicate just how much of a revitalizing effect the theory of plate tectonics has had on geology. Indeed, there is hardly any branch of the subject which has not benefited. Nearly two centuries after James Hutton we sense at last the formulation of a genuine Theory of the Earth. Although it is justification enough, the applications of the new concepts are not wholly academic. For obvious reasons there are very few explicit publications on the subject, but it is no secret that oil companies are taking full account of sea-floor spreading and associated phenomena in planning their exploration programmes.

What of the future? We are still too close to events to take any prognostication very seriously but there is no harm in trying.

It is not too much of a caricature to outline the progress of a new branch of science in the manner indicated in Fig. 45. Initially the payoff for one or two bright ideas, perhaps associated with the application of new techniques, is very considerable. A spectacular advance in knowledge is widely agreed to have taken place, probably after early scepticism. This we can term stage 1, which often carries with it the romantic image of a few imaginative men working in comparative isolation, intellectually speaking, in primitive conditions with the minimum of technical paraphernalia.

Many scientists are now attracted to the field and are kept busy developing and applying the new ideas and testing them with fresh evidence. A general atmosphere of excitement persists into this more mature phase, stage 2. As time passes the sense of novelty fades and the Law of Diminishing Returns (for a given effort) begins to operate with a vengeance. More and more research workers crowd into the field and the statisticians come into their own. This third stage approaches more

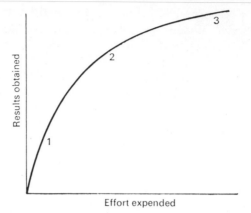

Results obtained

Effort expended

Fɪɢ. 45. The law of diminishing returns applied to phases of scientific research. For explanation see text.

to the layman's image of science, of huge research institutes with teams of sallow-faced technicians operating complicated machinery. A mountain of collective effort often seems to be rewarded by a molehill of a result. If this sounds too depressing, one should acknowledge that what Kuhn calls 'normal science' has many rewards for its practitioners, but the work is much too esoteric to capture the popular imagination. Moreover, one can rarely predict at what stage a new branch of science of a more specialized kind may sprout forth and renew the cycle. The opportunists of the scientific community are indeed constantly on the lookout for such developments.

My personal feeling is that plate tectonics has now moved into stage 2, though I would not like to predict when stage 3 will be reached. I suspect that it is still some distance off in time; certainly there is no shortage of challenging problems to tackle for quite a few years to come.

Postscript

Sɪɴᴄᴇ this book went to press, an important new idea has emerged from the fertile mind of Tuzo Wilson. Two types of mountain-building and subduction-zone formation are distinguished by considering the motion relative to the deeper mantle of two converging plates. First, a continental plate may be actively advancing and overriding an oceanic plate which is stationary relative to the mantle. In this case the subduction zone takes the form of coastal mountains as in Chile with a marginal trench which is pushed ahead of the continent. Second, an oceanic plate may be actively advancing and passing beneath a stationary continental plate. In this case island arcs and trenches form off shore as in East Asia, with little disturbance of the continental coast.

* Wɪʟsᴏɴ, J. T. *and* Bᴜʀᴋᴇ, K. (1972) *Nature, Lond.* **239,** 448.

8. Reflections on the revolution

The dispassionate intellect, the open mind, the unprejudiced observer, exist in an exact sense only in a sort of intellectualist folk-lore; states even approaching them cannot be reached without a moral and emotional effort most of us cannot or will not make.

WILFRED TROTTER

THE TITLE of this book contains an ambiguity. Did the 'revolution' in the Earth sciences commence with the formulation of the hypothesis of continental drift early this century or did it take place within the last few years, when it became more appropriate to speak of plate tectonics?

It is likely that many would subscribe to the latter view, and a case can certainly be made out for it. The *Oxford English Dictionary* defines a revolution as 'a complete change, a turning upside down, a fundamental reconstruction'. We have certainly had something close to that very recently in our views on Earth history. Ziman has shown that progress in a given field of science is marked by a change in consensus of the scientific community. In other words, a majority of what Ziman calls the 'invisible college' is converted from an old to a new set of beliefs about a particular phenomenon.

This has clearly happened since the mid 1960s in geology and geophysics. The change has been particularly dramatic in the United States. So strong was the feeling against continental drift until quite recently that in some institutions an open adherence to this doctrine would have put at serious risk the attainment of tenure by junior faculty members, while their more secure senior colleagues would have been all but drummed out of their invisible college. At the present time, only a few years later, almost the reverse situation holds. This surely has the flavour of revolution.[1] What is remarkable is not the existence of dissenters who are prepared to write articles or speak out at meetings, but their rarity. It would be unfortunate if all such people were to be dismissed as diehards. The more intelligent, less bigoted ones may well be able to point out anomalies in current theory which require resolution, or to indicate fields where plate tectonics fails to supply a satis-

---

[1] European scientists have tended to be less committal in their views either because they were more open-minded (the generous interpretation) or more cynical and indifferent. While on the subject of national differences, it is interesting to note that, although the propounder of continental drift was German, major post-war developments in the subject have been almost an Anglo-American monopoly.

factory explanation. Just such activities are part of the mopping up operations of Kuhn's 'normal science'. It is hard to conceive, however, of new evidence coming to light which would shake plate tectonic theory to its foundations, since so far it has passed all its major tests with flying colours.

It is of some interest to take brief note of some of the leading dissenters. Sir Harold Jeffreys has resolutely discounted the new geophysical and oceanographic evidence, as he had the geological and biological evidence earlier, preferring to place more reliance on, for example, his estimates of the viscosity of the mantle, which seems to him still to preclude significant lateral mobility. The Russian structural geologist V. Belousov has held in his mind a view of the Earth by which tectonic movements are predominantly vertical, and in which continental crust can somehow be converted into oceanic, though the evidence cited for this remarkable process has managed to convince few others. As with Jeffreys, his conceptual model of the Earth, to which he has been committed for many years, appears to have blinkered his attitude towards the new evidence. Warren Carey, though of course accepting the lateral migration of continents, still adheres to his model of a rapidly expanding Earth in the face of adverse evidence and hence rejects plate tectonics.

The American stratigrapher A. A. Meyerhoff falls in a different category from the others. He has somewhat heroically undertaken to challenge a large proportion of the evidence for continental drift, from the distribution of ancient evaporites and tillites to palaeomagnetic data, in a series of lengthy, belligerent articles which, if nothing else, attest to an amazing capacity to cope with a vast literature. Meyerhoff must be credited with the only sustained and serious effort to tackle the relevant evidence, but so far he has not succeeded in raising serious doubts among the Earth sciences community at large.

Returning to the subject of scientific revolutions, there is another interpretation of them which leads to a rather different conclusion. I refer to the influential work of T. S. Kuhn. Kuhn challenges the traditional view of scientific progress by the gradual accumulation of discoveries and inventions. Rather, revolutions occur by the replacement of one *paradigm* or world view by another. A paradigm is an accepted model or pattern of beliefs but is easier to explain by example than define. Kuhn draws his examples from physics and chemistry.

Take the case of optics, for instance. The universally held paradigm since early this century has interpreted light as photons. Previously light was treated wholly as wave motion and before that as corpuscles. Before the time of Newton there were a number of competing schools of thought, exhibiting a whole variety of incompatible ideas and *ad hoc* elaborations. Though the practitioners were genuine scientists the net result of all their activity was something less than science, for there was no sure foundation on which to build. Kuhn points out that the study of electricity was in a similarly confused state in the eighteenth century; major scientific progress was not made until the time of Franklin.

The acquisition of a paradigm is a sign of maturity in the development of a given field of science. In its absence, fact-gathering is more random and the weighting of evidence minimal. The mere accumulation of data in these circumstances can produce a veritable morass. In Francis Bacon's percipient words, 'Truth emerges more readily from error than from confusion.'

Turning back to the Earth sciences, it is quite clear that plate tectonics is the currently held paradigm.[1] To pinpoint the revolution in Kuhn's terms we must try to determine the paradigm which it has replaced, and this is more difficult.

Consider for instance the situation with regard to mountain-building. Earlier in this century, several totally different and mutually incompatible views have been widely held by those who refused to accept continental drift. One school saw crinkling of the crust through Earth contraction as the prime cause.[2] Others invoked local compression in a 'tectogene' by convergent and downward subcrustal convection currents, although where the currents came from was not specified. A third group denied the importance of lateral compression and instead invoked vertical uplift of narrow crustal sectors followed by lateral sliding under gravity. No convincing attempt was made, moreover, subsequent to the effort of Suess, to integrate orogenic phenomena with other facets of Earth history, such as marine transgressions and regressions, igneous activity, and changing climates. The various explanations of different phenomena thus had an *ad hoc* character. There was no consensus about which evidence was critical in deciding between rival hypotheses, if indeed there were any, and in consequence debate was frequently acrimonious.

Comparison is natural with pre-Newtonian optics as described by Kuhn, and confirms one's feeling that geology has become a more mature science in the last few years.

About the only common element that one can extract from such a

[1] One ought to note that a number of subjects within the broad field of Earth science, such as sedimentology and geochemistry, have made rapid progress recently without reference to whether continental drift has occurred or not.

[2] One would have thought that mountain-formation by Earth contraction was long discredited, but it was still being advocated, with full mathematical backing, as recently as the mid 1960s by a distinguished Cambridge astronomer, R. A. Lyttleton. No reference was made to the new developments in his paper in *Nature,* and it may be significant that he is a Fellow of the same college as Sir Harold Jeffreys, who has stubbornly adhered to his stabilist views.

hotchpotch of views is a belief in a 'stabilist' rather than a 'mobilist' Earth, that the continents have remained fixed in position with respect to each other. Since this was first seriously challenged by Wegener, and to a lesser extent by Taylor, the revolution can surely be considered to have started very early this century. Some fifty years were to pass before new evidence and ideas were to effect a mass conversion to the mobilist view and allow the full formulation of the new paradigm.

Half a century may seem a long period of indecision and uncertainty but it is not so very unusual in science, if one thinks, for instance, of Copernicus' heliocentric theory and Galileo's persecution for the same idea quite a long time later. Gillispie has excellently described the long conflict in early nineteenth-century geology between the catastrophists and uniformitarians. This was resolved in favour of the latter only by widespread acceptance of Lyell's *Principles of Geology*, although the key arguments and evidence presented by Hutton and Playfair had been available in the public domain for several decades. The dispute about whether or not large parts of Europe and North America had been covered by ice-caps in the recent geological past likewise persisted for several decades in the nineteenth century. There was in fact no total conversion, and the dispute only came to a close with the death of the last adherent to the old school. Charlesworth's absorbing account of the conflict makes one reflect on the blithe acceptance today of European Pleistocene glaciation as fact, whereas strictly speaking of course it is only a highly probable inference. Evolution serves as a third example. Although Darwin's hypothesis won over a consensus of biologists in a comparatively short time, the idea of evolution had been in the air since the beginning of the nineteenth century, and Lamarck's at least was a serious hypothesis.

Kuhn has been criticized for not defining the concept of the paradigm more clearly and for overdramatizing the contrast between 'revolutions' and 'normal science'. Certainly some might validly object to revolutions that persist unresolved for half a century *pari passu* with other more modest rates of change in thought and technique. It seems to me, however, that Kuhn has highlighted major features of science with a most illuminating conceptual model and has been perceptive in challenging the conventional view of cumulative progress. The Earth sciences do indeed appear to have undergone a revolution in the Kuhnian sense and we should not be misled by the fact that, viewed in detail, the picture may appear somewhat blurred at the edges.

Let us then turn back to Wegener and attempt, with the benefit of hindsight, to assess his work and the reaction to it.

It is not difficult to point out weaknesses. On specific points, such as the matching of moraines to prove continental linkage, Wegener was naïve. Biological evidence was used indiscriminately, regardless of whether or not the organisms in question had fossil records. The continental fits portrayed in his diagrams were extremely crude and hardly of the sort to convince a sceptic. An undue faith was placed in geodetic measurements, the proposed mechanism of drift was inadequate; and so on. But these failings pale into insignificance compared with his strengths.

The analysis of the scant geophysical data available at the time is masterly, as is the way he systematically pointed out inconsistencies and contradictions in conventional notions. He exhibited a clear logical mind and an ability to marshal relevant data from a wide variety of research fields. Penetrating insight into critical problems was shared with a refreshing freedom from the shackles of conventional wisdom.

Wegener showed a ready capacity to test his ideas. This is clearly brought out by his analogy of tearing a newspaper and finding that the print reads across when the severed parts are put together again. It is also evident in his checking of data based on the continental fit and matching the geology by reference to data on ancient climates. (With Köppen he ranks as the founder of the subscience of palaeoclimatology.) He was a staunch advocate of the multidisciplinary approach in Earth science. Since one cannot so readily devise critical experiments to test hypotheses as in physics, recourse must be had to the cumulative impact of evidence from a variety of fields, none of which may be decisive by itself. This is indeed a thoroughly respectable and in fact often the only way to proceed.

Why then did he fail to achieve a consensus in favour of his views? A number of reasons can be put forward, not all of them redounding greatly to the credit of the Earth science community.

One difficulty was certainly the lack of evidence, especially from the oceans, which in one way or another evidently held the key to interpretation, as shown so strikingly in the last decade. Some of the most critical areas for study, such as South Africa and South America, were effectively out of reach of the majority of geologists at a time when travel grants were a good deal less forthcoming than today. A lot, therefore, had to be taken on trust. This is insufficient reason, though, for dismissing continental drift out of hand; a more reasonable attitude and one adopted by a not insignificant number at least in Europe would have been fence-sitting reservation of judgement until more decisive data came to hand. One might go further and suggest that the stumbling-block was not so much the inadequacy of data as the stabilist paradigm or *Gestalt* of the Earth. Mere facts such as the shape of continents can be equivocal according to whether the stabilist or mobilist viewpoint is adopted.

It was widely accepted that the manifest inadequacy of Wegener's proposed mechanism was perhaps the single most important obstacle to acceptance. Yet gravity, geomagnetism, and electricity were all fully accepted long before they were adequately 'explained'. It could be argued of course that one would be foolish to deny phenomena that could manifest themselves to the practical man in various ways, whereas continental migration seems a much more nebulous affair making no direct impact upon the senses. There are better analogies from sciences associated with natural history.

The existence of former ice ages, notably in the Pleistocene, is universally accepted but there is no general agreement about the underlying cause. Perhaps the example of evolution is even more instructive. In the first forty years or so after *The origin of species* was published it became increasingly evident that the kinds of organic variation on which Darwin had pinned his faith were insufficiently heritable to provide a sound basis for natural selection. This did not deter biologists from believing in the *facts* of evolution. After Mendelian mutations were discovered at the turn of the century the opposite difficulty arose: the variations seemed too sharp for gradual adaptation. Yet acceptance of evolution as the paradigm persisted unshaken though another three decades were to pass before the difficulties were overcome by 'the new synthesis'. This case is illuminating because in truth large-scale organic evolution is as much an inference as continental drift.

It is perhaps more to the point that the geophysical arguments against drift put forward by Jeffreys were backed up by quantitative observations and an apparently superior knowledge of the physical properties of the Earth. One cannot fail to be struck by the comparison with the dispute concerning the age of the Earth between Lord Kelvin and the geologists of late last century, in which the high prestige of the physicist held sway until his assumptions were undermined by the discovery of radioactivity in rocks.

Kelvin set himself the problem in 1862 of calculating the time that had elapsed since the Earth had consolidated from a molten state and newly born from the sun. His initial time-limits were very wide because of uncertainty in the data, and ranged between 20 and 400 million years ago. By 1897 he felt justified in narrowing the limits to between 20 and 40 million years, which provoked vigorous protests from leading geologists of the time, such as Archibald Geikie.

By and large the geologists were frustrated in their protests because they were unable to counter the great physicist's calculations and assumptions in any effective quantitative or theoretically rigorous way. Perhaps they were cowed by Kelvin's famous but rather pernicious dictum, 'If you cannot measure, your knowledge is meagre and unsatisfactory.' In retrospect, as Holmes pointed out, it is easy to see the basic flaw in Kelvin's argument, that the Earth ought to be cooling because it is losing heat (we would not make the same assumption about an electric fire). The critical discovery which undermined Kelvin's case was made by another distinguished physicist who later became Lord Rayleigh. Rayleigh reported in 1906 on the detection of radium in a wide variety of rocks from all over the Earth. Possible moral: set a physicist to catch a physicist; they understand each other's language.

A fuller analysis of the difficulty faced by lack of a plausible mechanism for drift demands an understanding of the role of conceptual models in scientific thinking. Harré has distinguished two major categories of model, *homoeomorphic*, in which the subject of the model is the same as the source, as in a toy car, and *paramorphic*, in which the subject and source differ. Formulation and development of the paramorphic model, which requires an effort of creative imagination, is not quite the irrational process sometimes portrayed. It is subject to what Harré terms distraints, constraints, and restraints. *Distraints* represent the discipline exercised by observation, *constraints* the requirement that the hypothesis be intelligible, and *restraints* that it be plausible. Intelligibility and plausibility are both subject to the generally unnoticed groundswell of scientific thought, to the inexplicit assumptions and predelictions of the contemporary scientific community. The disciplined imagination, guided by analogy, conceives of a particular picture and awaits the development of suitable techniques to make that picture manifest. In the absence of a plausible mechanism for continental drift few Earth scientists could be persuaded to abandon their stabilist *Gestalt*, insecurely based though it was in terms of sheer factual support. The shortcoming was not to be properly made up until several decades after Wegener's hypothesis had become widely known.

Another reason for lack of acceptance was the partiality of Wegener's critics, who were normally only concerned with their own specialized field. It is quite clear reading the inter-war literature that there was virtually no intellectual communication, for example, between the geophysicists on the one hand and the biologists (including palaeontologists) on the other. It required an outsider such as Wegener to point out that the latter group's land bridges were untenable, but that the intercontinental faunal and floral similarities had to be explained somehow.

Consider the following quotation from Chapter 6 of Wegener's book, in which he berates those who adhered to their belief in transoceanic land bridges.

. . . a large proportion of today's biologists believe that it is immaterial whether one assumes sunken continental bridges or drift of continents—a perfectly preposterous attitude. Without any blind acceptance of unfamiliar ideas, it is possible for biologists to realise for themselves that the earth's crust must be made of less dense material than the core, and that, as a result, if the ocean floors were sunken continents and thus had the same thickness of lighter crustal materials as the continents, then gravity measurements over the oceans would have to indicate the deficit in attractive force of rock layer 4–5 km thick. Furthermore, from the fact that this is not the case, but that just about the ordinary values of gravitational attraction obtain over oceanic areas biologists must be able to form the conclusion that the assumption of sunken continents should be restricted to continental shelf regions and coastal waters generally, but excluded when considering the large ocean basins.

The message does not seem to have got across very clearly to the palaeobiogeographers, although Schuchert in 1932 did make some attempt to pare down the area of the land bridges required. Meanwhile the geophysicists seem to have been largely dismissive of the biological evidence, either because it was insufficiently quantitative to be taken seriously, or because they failed to understand the arguments.

The quotations given in Chapter 3, and others in similar vein challenging Wegener's credentials as a scientist, have a rather hollow ring today and seem to betray a failure to comprehend what scientific progress is all about.

Two views of the nature of scientific thought have held sway at different times. The more traditional view, associated especially with Francis Bacon and John Stuart Mill, is that it is essentially inductive, that one generalizes in some way from facts to theories. Many geologists in the past have adhered implicitly to Baconian principles. If only one collected facts assiduously enough they would eventually all fall into place. We are on the contrary more likely to be swamped by meaningless data. What is to guide our fact collecting, since observing everything is an impossibility? Piling up facts does not tell us why things behave as they do.

Medawar has attacked in withering terms the notion that science is essentially concerned with the collecting and classification of facts. On the contrary, the factual burden of a science varies inversely with its degree of maturity. As a science progresses particular facts become increasingly comprehensible within general statements of greater explanatory power and scope. In Sir Peter Medawar's words, we need no longer record the fall of every apple. The currently popular alternative view, lucidly expounded for instance by Medawar, has it that science is *hypothetico-deductive* in character. One points out flaws in conventional theory, hypothesizes an alternative interpretation on the basis of existing data, and deduces therefrom testable consequences.

This certainly gives a much more realistic account of the nature of scientific advance, and corresponds closely to what Wegener actually did: so much for his purported shortcomings as a scientist. Nevertheless the *hypothetico-deductive* model is still considered by Harré to give an inadequate and distorted account of how scientific progress is made.

Harré objects strongly to the rigid application of strictly logical canons to the question, as in Popper's 'falsification theory'. Evidence favourable to a particular hypothesis can be fraudulent, because true facts can be deduced from both true and false theories. As for unfavourable evidence, a knowledge of how scientists actually respond does not support the idea that this necessarily falsifies a given hypothesis. If science is nothing but a falsification of existing hypotheses by unfavourable evidence, how do we know how to replace them? Is it merely a random matter of trial and error?

The true picture is obscurer but more interesting, being bound up with the alteration of *Gestalte* or paradigms by the formulation of conceptual models subject to various restraints, and associated with all kinds of social factors. Viewed in these terms Wegener's role still appears very distinguished.

The trouble must partly have been that Wegener was not an accredited member of the professional geologists' club; at least one senses the undertones in the literature. We of course now see it as a positive advantage that Wegener had not been brainwashed by the conventional geological wisdom as a student. The key role that an outsider can play in transforming the world view in a particular subject is now well recognized. John Dalton is a good case in point, for he was not a chemist but, curiously enough like Wegener, a meteorologist. His main interest was in the physical problems of the absorption of gases by water and of water by the atmosphere. His whole approach, attitude, and training were quite different from those of contemporary chemists, and it led him to formulate his celebrated atomic theory which, needless to say, was widely attacked when first announced.

Finally, one must not underestimate the role of sheer conservative prejudice and intellectual fashion, and the greater emotional security obtained by adhering to established ideas. This seems indicated by the continued resistance to continental drift even after Holmes and du Toit between them had seemed to dispose of many of the original objections. The game is given away by Chamberlin in his article in the 1928 A.A.P.G. symposium, as he quotes with evident approval a remark made by a colleague, 'If we are to believe Wegener's hypothesis we must forget everything which has been learned in the last 70 years and start all over again.'

We should not be too surprised by all this. One may quote, for example, Beveridge, writing of science in general.

In nearly all matters the human mind has a strong tendency to judge in the light of its own experience, knowledge and prejudices rather than on the evidence presented. Thus new ideas are judged in the light of prevailing beliefs. If the ideas are too revolutionary, that is to say, if they depart too far from reigning theories and cannot be fitted into the current body of knowledge, they will not be acceptable. When discoveries are made before their time they are almost certain to be ignored or meet with opposition which is too strong to overcome, so in most instances they might as well not have been made.

The theory of plate tectonics, an outgrowth of the continental drift hypothesis, has proved highly successful in integrating diverse geological phenomena and contributing towards a more coherent and intelligible picture of the Earth's evolution than obtained hitherto. It must rank as the biggest advance in Earth science since acceptance in the early nineteenth century of the paradigms of uniformitarianism and stratigraphical correlation by fossils established geology as a true science. By the criteria normally used to judge the quality of scientific theories, namely precision, scope, explanatory value, and testability, plate tectonics scores highly. As the man who really started it all, Alfred Wegener deserves wider recognition as one of the most important scientific innovators of this century.

Appendix 1

The Phanerozoic time-scale[1]

ERA	PERIOD		AGE OF BASE (in millions of years)
Cainozoic	Quaternary	Pleistocene	2
	Tertiary	Pliocene	7
		Miocene	26
		Oligocene	38
		Eocene	54
		Palaeocene	65
Mesozoic		Cretaceous	136
		Jurassic	190
		Triassic	225
Palaeozoic		Permian	280
		Carboniferous	345
		Devonian	395
		Silurian	440
		Ordovician	500
		Cambrian	570

[1] Based on the time-scale published by the Geological Society of London in 1964.

Appendix 2

The World Rift System and Plate Tectonics, or 1971 and All That

BY

B. C. and G. C. P. King

'I'll put a girdle round about the earth in forty minutes' (Puck)

They put a girdle round the Earth
   And named it the Worldwide Rift;
It helps explain the ocean floor
   And Continental Drift.

Vine and Matthews sailed away
   Exploring the ocean bed;
It took much longer getting them back;
   They said it was seafloor spread.

It appears that the oceans were mostly formed
   By Cainozoic streams
Of mantle flooding up the cracks
   And gumming up the seams.

Our Earth can scarce make up its mind;
   It flips its magnetic poles
And the magnetized basalts so produced
   Give temporal controls.

When all believed the Earth was flat
   It was but a single plate,
But even then the edges were
   Hot subjects for debate.

They can't curl down; they must curl up
   To form a kind of dish
To stop the oceans spilling out
   And losing all the fish.

We now all know the Earth is round
   And moves about the Sun,
So what the ocean spreaders said
   Could not be simply done.

Unless the Earth itself enlarged
   And kept on getting bigger,
But gravity opposes this:
   It is a constant figure.

And so the plate has been revived
   In present day tectonics,
Though sial and sima still remain
   As crustal term mnemonics.

Both kinds of crust now constitute
   The grander types of plates
And as they move upon the Earth
   They suffer subtle fates.

The edges which are growing still
   Are hid beneath the oceans,
While those around the island arcs
   Show self-consuming motions.

Yet others seem to hit or slide
   Performing curious functions,
And where they can't make up their minds
   You there have triple junctions.

On continents the crustal plates
   Are edged by earthquake foci,
And little plates proliferate
   By joining up the loci.

As alchemists once sought the stone
   For magic transmutations,
The motions of the plates are shown
   By seismic computations.

Geologists naively thought
   That rifts were due to faulting,
By subsidence of crustal strips
   Along a pre-rift vaulting.

Their evidence was solely based
   On visual observations
Of structure and stratigraphy
   And such out-moded notions.

But others now hold better views
   And think that each 'mañana'
Brings Africa a step more close
   To the fate of old Gondwana.

McKenzie sees his moving plates
  Wedging the rift asunder,
And one day ships will sail the rift
  To maritime Uganda.

Girdler and Khan from the gravity highs
  That they have in the Gregory rift
See the mantle rearing its ugly head
  And East Africa going adrift.

But the OAU no doubt will vote
  At a suitable early date
To stop that drift and plug the rift
  Before it is too late.

Addendum: Headline in *The Times*, 18 June 1971, page 6: 'Emperor tries to close African rift.'

References and further reading

CHAPTER I

CAROZZI, A. V. (1970) A propos de l'origine de la théorie des dérives continentales: Francis Bacon (1620), François Placet (1668), A. von Humboldt (1801) et A. Snider (1858). *C. R. Séances Soc. Phys. Hist. nat. Geneva*, N. S. **4,** 171–9.

DU TOIT, A. L. (1937) *Our wandering continents.* Oliver and Boyd, Edinburgh.

HARLAND, W. B. (1969) The origin of continents and oceans: an essay review. *Geol. Mag.* **106,** 100–4.

RUPKE, N. A. (1970) Continental drift before 1900. *Nature, Lond.* **227,** 349–350.

SNIDER-PELLEGRINI, A. (1858) *La Création et ses mystères dévoilés.* Franck and Dentu, Paris.

TAYLOR, F. B. (1910) Bearing of the Tertiary mountain belt on the origin of the Earth's plan. *Bull geol. Soc. Amer.* **21,** 179–226.

CHAPTER 2

GEORGI, J. (1962) Memories of Alfred Wegener. *In* S. K. Runcorn (editor): *Continental drift.* Academic Press, New York.

SUESS, E. (1904–9) *The face of the Earth* (5 vols.) Clarendon Press, Oxford.

WEGENER, A. (1912) Die Entstehung der Kontinente. *Petermanns Mitteilungen,* 185–95, 253–6, 305–9.

— (1966) *The origin of continents and oceans.* Translated from the 4th revised German edition of 1929 by J. Biram, with an introduction by B. C. King. Methuen, London.

CHAPTER 3

ARGAND, E. (1924) La tectonique de l'Asie. *C. R. XIII Congr. géol. int.* 1922, Liège, 171–372.

COLEMAN, A. P. (1925) Permo–Carboniferous glaciation and the Wegener hypothesis. *Nature, Lond.* **115,** 602.

DALY, R. A. (1926) *Our mobile Earth.* Scribner, New York.

DU TOIT, A. L. (1927) A geological comparison of South America with South Africa. *Carnegie Inst. Wash. Publ.* **381,** 1–157.

— (1937) *Our wandering continents.* Oliver and Boyd, Edinburgh.

HOLMES, A. (1929) Radioactivity and earth movements. *Trans. geol. Soc. Glasgow* **18,** 559–606.

JEFFREYS, H. (1924, 1929) *The Earth: its origin, history and physical constitution.* (1st and 2nd editions) Cambridge University Press.

LAKE, P. (1922) Wegener's displacement theory. *Geol. Mag.* **59,** 338–46.

VAN WATERSCHOOT VAN DER GRACHT, W. A. J. M., *et al.* (1928) *Theory of continental drift: a symposium.* American Association of Petroleum Geologists, Tulsa.

WASHINGTON, H. S. (1923) Comagmatic regions and the Wegener hypothesis. *J. Wash. Acad. Sci.* **13,** 339–47.

WRIGHT, W. B. (1923) The Wegener hypothesis. *Nature, Lond.* **111,** 30–1.

CHAPTER 4

BRIDEN, J. C., SMITH, A. G. *and* SALLOMY, J. T. (1970) The geomagnetic field in Permo-Triassic time. *Geophys. J. R. astr. Soc.* **23,** 101–17.

CAREY, S. W. (1958) A tectonic approach to continental drift. *Geol. Dept. Univ. Tasmania, Sympos. no. 5,* 177–355.

GOLD, T. (1955) Instability of the Earth's axis of rotation. *Nature, Lond.* **175,** 526–9.

HEEZEN, B. C. (1962) The deep sea floor. *In* S. K. Runcorn (editor): *Continental drift.* Academic Press, New York.

HILL, M. N. (editor) (1963) *The sea;* Vol. 3, *The Earth beneath the sea.* Wiley-Interscience, New York.

IRVING, E. (1964) *Paleomagnetism and its application to geological and geophysical problems.* Wiley, New York.

MCELHINNY, M. W. *and* BRIDEN, J. C. (1971) Continental drift during the Palaeozoic. *Earth Planet. Sci. Letters* **10,** 407–16.

MENARD, H. W. (1964) *Marine geology of the Pacific.* McGraw Hill, New York.

RUNCORN, S. K. (1962) Palaeomagnetic evidence for continental drift and its geophysical cause. *In* S. K. Runcorn (editor): *Continental drift.* Academic Press, New York.

TARLING, D. H. (1972) *Principles and applications of palaeomagnetism.* Chapman and Hall, London.

VACQUIER, V. (1962) Magnetic evidence for horizontal displacement in the floor of the Pacific Ocean. *In* S. K. Runcorn (editor): *Continental drift,* Academic Press, New York.

CHAPTER 5

COX, A., DALRYMPLE, G. B. *and* DOELL, R. R. (1967) Reversals of the Earth's magnetic field. *Sci. Amer.* **216,** 44–54.

DIETZ, R. S. (1961) Continent and ocean basin evolution by spreading of the sea floor. *Nature, Lond.* **190,** 854–7.

HEIRTZLER, J. R., *et al.* (1968) Marine magnetic anomalies, geomagnetic field reversals and motions of the ocean floor and continents. *J. geophys. Res.* **73,** 2119–36.

HESS, H. H. (1962) History of ocean basins. *In* A. E. J. Engel *et al.* (editors): *Petrologic studies: a volume in honor of A. F. Buddington.* Geological Society of America, Boulder, Colorado.

HOLMES, A. (1929) Radioactivity and earth movements. *Trans. geol. Soc. Glasgow* **18,** 559–606.

ISACKS, B., OLIVER, J. *and* SYKES, L. R. (1968) Seismology and the new global tectonics. *J. geophys. Res.* **73,** 5855–99.

MAXWELL, A. E. (editor) (1971) *The sea*; Vol. 4, *New concepts of sea floor evolution*. Wiley-Interscience, New York.

— *et al.* (1970) Deep sea drilling in the South Atlantic. *Science* **168,** 1047–59.

VINE, F. J. (1966) Spreading of the ocean floor: new evidence. *Science* **154,** 1405–15.

— (1971) Sea floor spreading. *In* I. G. Gass *et al.* (editors): *Understanding the Earth*. Published for the Open University Press by Artemis Press, 'Sussex.

— *and* MATTHEWS, D. H. (1963) Magnetic anomalies over oceanic ridges. *Nature, Lond.* **199,** 947–9.

WILSON, J. T. (1965) A new class of faults and their bearing on continental drift. *Nature, Lond.* **207,** 343–7.

CHAPTER 6

DEWEY, J. F. (1972) Plate tectonics. *Sci. Amer.* **226** (May), 56–66.

ISACKS, B., OLIVER, J. *and* SYKES, L. R. (1968) Seismology and the new global tectonics. *J. geophys. Res.* **73,** 5855–99.

LE PICHON, X. (1968) Sea floor spreading and continental drift. *J. geophys. Res.* **73,** 3661–97.

MCKENZIE, D. P. (1969) Speculations on the consequences and causes of plate motions. *Geophys. J. roy. astr. Soc.* **18,** 1–32.

— *and* PARKER, R. L. (1967) The North Pacific: an example of tectonics on a sphere. *Nature, Lond.* **216,** 1276–9.

MORGAN, W. J. (1968) Rises, trenches, great faults and crustal blocks. *J. geophys. Res.* **73,** 1959–82.

— (1972) Deep mantle convection plumes and plate motions. *Bull. Am. Ass. Petrol. Geol.* **56,** 203–13.

— *and* MCKENZIE, D. P. (1969) Evolution of triple junctions. *Nature, Lond.* **224,** 125–33.

OXBURGH, E. R. (1971) Plate tectonics. *In* I. G. Gass *et al.* (editors): *Understanding the Earth*. Published for the Open University by Artemis Press, Sussex.

CHAPTER 7

BULLARD, E. C., EVERETT, J. E. *and* SMITH, A. G. (1965) The fit of the continents around the Atlantic. *Phil. Trans. roy. Soc. Lond.* A **258,** 41–51.

CROWELL, J. C. *and* FRAKES, L. A. (1970) Phanerozoic glaciation and the causes of ice ages. *Am. J. Sci.* **268,** 193–224.

DEWEY, J. F. (1971) A model for the Lower Palaeozoic evolution of the southern margin of the early Caledonides of Scotland and Ireland. *Scot. J. Geol.* **7,** 219–40.

— *and* BIRD, J. (1970) Mountain belts and the new global tectonics. *J. Geophys. Res.* **75,** 2625–47.

DICKINSON, W. R. (1970) Relations of andesites, granites and derivative sandstones to arc–trench tectonics. *Rev. Geophys. space Phys.* **8,** 813–60.

— (1971) Plate tectonics in geologic history. *Science* **174,** 107–13.

FALCON, N. L., *et al.* (1970) A discussion on the structure and evolution of the Red Sea and the nature of the Red Sea, Gulf of Aden and Ethiopia rift junction. *Phil. Trans. roy. Soc. Lond.* A **267,** 1–412.

HALLAM, A. (1971) Mesozoic geology and the opening of the North Atlantic. *J. Geol.* **79,** 129–57.

— (1972) Continental drift and the fossil record. *Sci Amer.* **227** (Nov.), 55–66.

— (editor) (1973) *Atlas of palaeobiogeography.* Elsevier, Amsterdam.

HUGHES, N. F. (editor) (1972) Faunas and continents through time. *Palaeont. Ass. Spec. Publ.* **12.**

KEAST, A. (1971) Continental drift and the evolution of the biota on southern continents. *Quart. Rev. Biol.* **46,** 335–78.

LAUGHTON, A. S. (1971) South Labrador Sea and the evolution of the North Atlantic. *Nature, Lond.* **232,** 612–17.

McKENZIE, D. P. *and* SCLATER, J. G. (1971) The evolution of the Indian Ocean since the late Cretaceous. *Geophys. J.* **24,** 437–528.

MITCHELL, A. H. *and* READING, H. G. (1971) Evolution of island arcs *J. Geol.* **79,** 253–84.

PITMAN, W. C. *and* TALWANI, M. (1972) Sea floor spreading in the North Atlantic. *Bull. geol. Soc. Am.* **83,** 619–46.

SCLATER, J. G. *and* FRANCHETEAU, J. (1970) The implications of terrestrial heat flow observations on current tectonic and geochemical models of the crust and upper mantle of the Earth. *Geophys. J. roy. astr. Soc.* **20,** 509–42.

SMITH, A. G. (1971) Alpine deformation and the oceanic areas of the Tethys, Mediterranean and Atlantic. *Bull. geol. Soc. Am.* **82,** 2039–70.

—*and* HALLAM, A. (1970) The fit of the southern continents. *Nature, Lond.* **225,** 139–44.

VALENTINE, J. W. *and* MOORES, E. M. (1970) Plate tectonics regulation of faunal diversity and sea level: a model. *Nature, Lond.* **228,** 657–9.

VAN ANDEL, Tj. H. *and* HEATH, G. R. (1970) Tectonics of the Mid-Atlantic Ridge 6–8° south latitude. *Mar. geophys. Res.* **1,** 5–36.

VEEVERS, J. J., *et al.* (1971) Indo-Australian stratigraphy and the configuration and dispersal of Gondwanaland. *Nature, Lond.* **229,** 383–8.

CHAPTER 8

BEVERIDGE, W. I. B. (1950) *The art of scientific investigation.* Heinemann, London.

CHARLESWORTH, J. K. (1957) *The Quaternary Era, with special reference to its glaciation.* vol. 1. Arnold, London.

GILLISPIE, C. C. (1959) *Genesis and geology: the impact of scientific discoveries upon religious beliefs in the decades before Darwin.* Harper, New York.

HARRÉ, R. (1970) Constraints and restraints. *Metaphilosophy* **1,** 279–99.

— (1970) *The principles of scientific thinking.* MacMillan, London.

KUHN, T. S. (1962) *The structure of scientific revolutions.* Univ. of Chicago Press.

MEDAWAR, P. B. (1967) *The art of the soluble.* Methuen, London.

ZIMAN, J. M. (1968) *Public knowledge: an essay concerning the social dimension of science.* Cambridge University Press.

Index